Introduction to
Reversible Computing

Chapman & Hall/CRC
Computational Science Series

SERIES EDITOR

Horst Simon

Deputy Director

Lawrence Berkeley National Laboratory

Berkeley, California, U.S.A.

PUBLISHED TITLES

Introduction to
Reversible Computing

Kalyan S. Perumalla

Oak Ridge National Laboratory
Knoxville, Tennessee, USA

CRC Press
Taylor & Francis Group
Boca Raton London New York

CRC Press is an imprint of the
Taylor & Francis Group, an **informa** business

A CHAPMAN & HALL BOOK

CRC Press
Taylor & Francis Group
6000 Broken Sound Parkway NW, Suite 300
Boca Raton, FL 33487-2742

© 2014 by Taylor & Francis Group, LLC
CRC Press is an imprint of Taylor & Francis Group, an Informa business

No claim to original U.S. Government works

ISBN-13: 978-1-4398-7340-3 (hbk)

Visit the Taylor & Francis Web site at
http://www.taylorandfrancis.com

and the CRC Press Web site at
http://www.crcpress.com

Contents

List of Figures

List of Tables

List of Algorithms

Preface

The concept of reversible computing is based on a combination of forward and backward computation, in contrast to traditional forward-only computation. This book provides an introduction to the many facets of reversible computing, including theory, hardware and software aspects, fundamental limits, complexity analyses, practical algorithms, compilers, efficiency improvement techniques, and application areas. The topics covered here span several areas of computer science, including high-performance computing, parallel/distributed systems, computational theory, compilers, power-aware computing, and super-computing.

The book includes sufficiently basic material for readers new to reversible computing to easily get started. For those who are already familiar with a certain topic within reversible computing, the book can serve as a one-stop reference to other topics in reversible computing, providing an expanded view. The expanded view of reversible computing includes the traditional energy-motivated hardware views as well as the emerging application-motivated software views.

Although there are other books on engineering and logic design for reversible computing, few comprehensively cover the software and programming aspects. Filling this gap, new emphasis is introduced in this book on the software, programming, applications, and usage aspects in the expanded view of reversible computing.

The book itself is envisioned to be suitable at the senior undergraduate and graduate levels. However, it is also intended as a handbook for professionals in the industry and government research laboratories. Students, faculty, and professionals in computer science, computer engineering, applied physics, applied mathematics, and related disciplines would benefit from this book. The book may be used as supplementary reading in courses such as parallel and distributed systems, high-performance computing, parallel simulation, and computational physics.

In order to suit the broader readership of the book, simplicity was preferred over rigor, wherever the choice was warranted. Citations to original articles on seminal results are provided so that interested readers may consult the corresponding publications in the literature. Also, in chapters that cover advanced material, pointers to the additional resources are provided for further reading in specific sub-topics. No specific, conscious attempt is made to recount a historical perspective of development in reversible computing.

About the Author

Kalyan Perumalla, PhD, is a senior R&D staff and manager in the Computational Sciences and Engineering Division at the Oak Ridge National Laboratory, Oak Ridge, and an adjunct professor in the School of Computational Sciences and Engineering at the Georgia Institute of Technology, Atlanta. Dr. Perumalla also founded and currently leads the High Performance Discrete Computing Systems team in the Modeling and Simulation Group at the Oak Ridge National Laboratory. He earned his PhD in computer science from the Georgia Institute of Technology in 1999. His areas of interest include reversible computing, high-performance computing, parallel discrete event simulation, and parallel combinatorial optimization. Dr. Perumalla is a winner of the prestigious US Department of Energy Career Award in Advanced Scientific Computing Research, 2010–2015. His primary research contributions are in the application of reversible computation to high-performance computing and in advancing the vision of a new class of supercomputing applications using real-time, parallel discrete event simulations. In addition to this book, he has co-authored another book, three book chapters, and over 100 articles in peer-reviewed conferences and journals. Four of his co-authored papers received the best paper awards, in 1999, 2002, 2005 and 2008, and two were finalists in 2010.

Dr. Perumalla has been actively serving the research community in the roles of program committee member and reviewer for several international conferences and journals. He serves on the editorial board of the *ACM Transactions on Modeling and Computer Simulation* and the *SCS Transactions of the Society for Modeling and Simulation International*. His research prototype tools in parallel and distributed computing have been disseminated to research institutions worldwide. He has performed research as an investigator on several research programs sponsored by US federal agencies, including the Department of Energy, Department of Defense, Department of Homeland Security, and the National Science Foundation.

Acknowledgments

I would like to acknowledge the gracious reviewing efforts by Vladimir Protopopescu, James Nutaro, and Sudip Seal, who provided many useful comments and suggestions for improvement on the drafts of this book. The support and patience from the publisher, especially from Randi Cohen, are gratefully acknowledged. Regular bouts of warm encouragement and advice from Vladimir Protopopescu served to keep my spirits high for many professional goals, this book included. Finally, it would have obviously been impossible for the book to materialize without help from the family—I enjoyed the luxury of the most loving and considerate support from my wife and sons while I completed this book.

Book Organization

This book is organized into four parts: Introduction, Theory, Software, and Hardware.

In Part I, Chapter 1 describes the scope of reversible computing. Chapter 2 covers the range of application areas in which reversible computing finds uses. Chapter 3 presents the spectrum of implementation components that comprise the realization of reversible computing in practice.

Part II deals with theory, models, and paradigms underlying reversible computing. This part is aimed at providing a quick overview of some well-known theoretical results related to reversible computing. Some of the most fundamental results are described in Chapter 4 on the technology-independent facets of reversible computing, especially with system-theoretic views. Chapter 5 is devoted to the description and resolution of several related paradoxes. This includes illustrations of the commonly misunderstood distinction of thermodynamical entropy from memory for reversal, and also a treatment of the Maxwell's Demon, among others. In Chapter 6, a pivotal development is presented, namely the construction of a reversible Turing machine model that can simulate any conventional, irreversible Turing machine. A new categorization of paradigms for relaxing irreversible execution into reversible execution is described in Chapter 7.

Part III focuses on the software aspects of reversible computing, covering the higher-level computing elements, including programming languages, compilers, automated history reduction, random number generation, and numerical computation. Readers well versed in traditional forward programming will benefit from these chapters on reversible software by learning (1) why and how conventional software constructs are inadequate for reversible execution, (2) different ways in which reversibility can be introduced into software, and (3) techniques by which the efficiency of reversibility is increased in specific classes of codes. Chapter 8 is devoted to the relatively new area of reversible programming languages that provide reversibility *by design*, with articulation of the issues and requirements in reversible programming language constructs, and illustration of a few reversible languages. The issue of adding reversibility to conventional, irreversible programs is addressed in Chapter 9, which describes multiple checkpointing and backward computation schemes. For a practical implementation, the challenge of making arbitrary **C** programs reversible is studied in Chapter 10. In this chapter, a source-to-source methodology is described to convert any **C** program to an equivalent but reversible version that retains the original forward–only semantics. The potential for more efficient reversal by transcending a localized view of program statements is described in Chapter 11, with an algorithm for reversing a class of programs called linear codes. New, zero-memory methods for generating pseudo-random number streams in a reversible fashion are described in Chapter 12. The methods highlight the complexity of the reversibility problem and the solutions for use in reversible execution of scientific applications that rely on complex random number streams. New solutions to the problem of adding reversibility to dynamic memory allocation and deallocation operations are proposed in Chapter 13. The immense challenge of making numerical computation reversible is addressed in Chapter 14, identifying the sources of irreversibility and outlining the possible solution approaches. Automated ways for reversal relying on history are given for conventional interfaces of computer-based arithmetic. New, zero-memory solutions are also proposed for integer or fixed-point arithmetic based on novel internal bit representations and an abridged set of novel definitions of reversible arithmetic operations. In Chapter 15, the problem of reversing the operation of sorting algorithms in a memory-efficient way is described, along with three solution approaches. A generalized template for the implementation of the Undo–Redo–Do paradigm of reversible computing is presented in Chapter 16, which can be used in a wide range of user-interface applications that provide undo capabilities for user-initiated actions.

Part IV covers the hardware concepts in implementing reversible computing and the descriptions of some of their actual realizations. The circuit and gate-level reversibility is covered in Chapter 17 on reversible gates and their synthesis. Chapter 18, on the computer architectural designs, presents the low-level instruction sets for reversible machine languages and memory interface units.

The book concludes with Part V identifying some important research directions in the future of reversible computing.

Part I

Introduction

Chapter 1

Scope

1.1   Notions of Computing

Consider the word *computation*. For some strange and unknown reason, the notion that naturally arises in one's mind is that of *forward* computation. The unidirectionality of computational flow seems to be the natural and intuitive notion for many people. Even more interestingly, no notion of computation in the opposite direction seems to be naturally triggered in one's mind—neither a mental model nor even a recollection of any related reversible phenomenon. Nevertheless, technically savvy persons, when prodded by even a minor hint of the reversal possibilities, are able to quickly pursue the line of thought a bit further and soon uncover a rough idea of the immense challenges and limitations of reversible computation. Often, it seems to be easy to arrive at a quick (and hasty) conclusion that reversibility is impossible and consequently not worth contemplating further. The author had the occasion to bring the topic of reversible computing into conversations with people from a variety of backgrounds; in almost all cases, the notion was encountered with initial surprise, followed by brief disbelief, followed by a more sustained opinion that varied with the background of the person. For example, the simple response is usually one of "...but that is impossible, right...?" or "...but why would one ever use that?" The response from technically educated people (e.g., scientists from other domains) is "...but there are computations that are fundamentally irreversible, so it would not work..." or "...entropy always increases, so there is fundamentally no such thing as reversible," and so on.

Given that this has been the state of affairs with respect to knowledge about reversibility of computation, it is probably not surprising that reversible computing has experienced modest and slow progress. Admittedly, computing backwards is not at all common in daily life. In situations where it does appear in real life, the process is not easily understood (take, for example, mortgage escrow payment calculators, which require computing backward and forward).

Forward-only computing has enjoyed the benefits of enormous efforts in its research, development, and education all over the world, in academia, industry, and government. Electronic computing essentially started with the de-facto forward-only mode at least six decades ago (for the history of computation, see, for example, [Goldstine, 1980]), and has since been exercised in very many domains. Prior to that, mechanical computing, dating back to the 1800s, was also focused exclusively on the forward-only mode [Swade, 2001]. Thus, considering the product of the number of people and the length of time spent by the people (*man-years*), forward-only computing has already enjoyed several thousand man-years of research, even by conservative estimates. In contrast, as of this writing, all the researchers worldwide in reversible computing can indeed fit in a small auditorium, with room to spare. Relative to conventional forward-only computing, the know-how in reversible computing is currently limited. Nevertheless, some recent advances and a corresponding increase in the base of expertise in reversible computing have now reached noticeable levels, both of which experiencing a spurt in just the past two to three decades. There is now a sizable, albeit scattered, body of works in various disparate subareas within reversible computing. More importantly, reversibility is emerging as one of the most exciting new *dimensions* in computing for the future, positioned for inevitable progress and expansion in the coming decades.

1.2   A Whole New Dimension

In a way, it can be argued that reversible computing is to computing what imaginary numbers are to real numbers. Just as imaginary numbers subsume real numbers, reversible computing subsumes traditional computing, making it a special case, and generalizes it. The consideration of reversibility adds an entirely new, "orthogonal" dimension to almost all aspects of traditional computing. The "orthogonality" of this new dimension has profound and wide implications. The word "reversible" now becomes an adjective that can be used to qualify every traditional (forward-only) computing concept. For example, computer arithmetic gives two relaxations: conventional arithmetic and reversible arithmetic. Similarly, conventional dynamic memory management results in two relaxations: forward-only dynamic memory and reversible dynamic memory. For all concepts in computing, reversibility gives rise to a new, fertile class of open research problems.

1.3 Related Terms and Synonyms

Reversible computing means different things to different people. Together with the concept of reversible computing, several other concepts and related terms are encountered. A few of them are used somewhat synonymously, while other terms and phrases appear in relation to some specific aspect of reversible computing. Among the common ones are

■ *Reversible computing* This is the most recognized phrase for referring to the largest set of concepts related to computing in which backward execution is used. Although historically this has been associated with certain technologies aimed at minimizing the energy consumption of any computation, it has been the term of choice to refer to hardware as well as software aspects. In this book, by reversible computing, we refer to all concepts traditionally associated with the term, and also include other computing paradigms (see Chapter 7) within which reversible computing appears in various forms.

■ *Reversible logic* This is often used to refer to the low-level hardware aspects of reversible computing, namely the gate-level concepts such as reversible truth tables, reversible logic gates, and synthesis of reversible circuits.

■ *Adiabatic computing* Adiabatic computing is another term that has been often used to denote the hardware circuit technology whose energy consumption can be (ideally) fully recovered and reused. Note that the term itself is a misnomer (see, for example, Section 7.3 of [Frank, 1999]). While an *adiabatic* process involves zero heat transfer, an "adiabatic" circuit does not necessarily preclude heat transfer. Because it is entropy that is being conserved (suggesting the adjective *isentropic*), and even entropy can never be completely conserved unless the operating frequency is close to zero (suggesting the adverb *asymptotically*), a more appropriate alternative term for adiabatic computing would be *asymptotically isentropic* computing, as suggested in [Frank, 1999].

■ *Reverse execution* This term is used to refer to program-level reversibility, especially in relation to actual execution techniques for inverting the control flow. Almost invariably, this phrase conveys the software dimension of the execution, as opposed to the hardware aspects.

■ *Reverse computation* This is used synonymously with *reverse execution*, but perhaps with more implied emphasis on the data and arithmetic aspects of reversibility, in addition to the control flow.

■ *Invertible programming* This term typically alludes to reversibility at the programming language and software level of application. It implies a notion of built-in reversibility into the language constructs, which naturally results in programs that can be reversed.

■ *Inverse programming* This term typically alludes to reversibility at application-level, similar to that of *invertible programming*, except that the emphasis is more on the program and less on the programming language support.

■ *Bi-directional execution* This term applies to any program execution that carries with it the notion of a runtime choice between forward or backward execution, regardless of the granularity of the execution unit. Examples include support for undoing actions in word processing, and algorithms for physics simulations that can simulate particle motion forward or backward in time.

■ *Backward execution* This term, sometimes intended as general reversible execution, refers to the backward part of a bi-directional execution.

■ *Reversible execution* This term is used synonymously with bi-directional execution.

■ *Anti-computation* This term is used to signify the nullification aspect of reversible computing. By analogy to matter and anti-matter, when computation and anti-computation come together, they annihilate each other, leaving no trace of either. This notion brings with it the allowance for units of computation and anti-computation to be generated in arbitrary order, and for the annihilation to occur regardless of their histories.

■ *Injective programming* This term is used to highlight the functional view of reversible execution, namely that the function defined by the computation (input-to-output mapping) is required to be injective (one-to-one). Sometimes the term *bijective programming* is also used to denote those injective mappings that are also *onto*, ensuring that every value belonging to the range (output) also has a uniquely mapped value in the domain (input). Thus, in bijective programming, one could start with any output value and work backward toward its corresponding input value. In injective programming, this is not always possible.

■ *Others* Other related terms include *speculative execution*, *optimistic execution*, and *quantum computing*, although each of these terms is specific to a particular implementation approach or usage of reversible computing.

1.4 Similar yet Unrelated Concepts

There are a few other concepts that are different from reversible computing but often confused with it.

■ *Reverse Engineering* This is a concept that appears in software engineering. Reverse Engineering is concerned with the art of deciphering a previous form of a software artifact before it was transformed or converted to the current form. The most common example is the process of deciphering the source code from the executable object code, the former being unavailable (e.g., due to proprietary concerns), but the latter being available. Because the compiler transforms the source code to object code in very complex ways, it is not an easy task to arrive at the most likely source code that produced a given object code. Another example is the uncovering of the pseudocode of an algorithm from its expression in source code or object code, especially when the code is intentionally obfuscated. Tools for reverse engineering are typically based on heuristics with domain know-how, and do not provide general solutions. Although notionally there is a conceptual element of reversal in the action, this is not reversible computing.

■ *Backtracking* This is a concept appearing in logic programming in which multiple paths are explored in a backward mode in an attempt to find the path that best matches a given outcome. Although it is not the same as reversible computing, backtracking could be an algorithmic approach in solving some reversible computing problems such as in finding a way to trade-off computation time for storage so that past state(s) can be recreated from current state by exploring past states that best match the current state.

■ *Inverse Methods* There is a range of numerical methods used in complex scientific simulations to determine the initial conditions that lead to specific final conditions. The simulations use numerical models of complex phenomena, typically described by coupled partial differential equations, and the inversion requires a combination of analysis and numerical inversion to arrive at the initial conditions of interest. Again, there is an element of reversal involved in inverse methods, but the context is very specific to the numerical methods.

■ *Encryption and Decryption* Cryptography is related to reversible computing to the extent that encryption and decryption are reversals of each other. However, encryption and decryption only happen to be just one of the forward-reverse pairs of functions in the vast, general set of functions in reversible computing.

Chapter 2

Application Areas

2.1 General Reversible Computing Problem

At a high level, the challenge of reversible computing may be stated as follows: Given a program, the program must be executed in such a way that its forward progress could be paused at any moment and the direction of its execution can be reversed to retrace its previous steps exactly. The backward execution can also be paused at any moment and the forward execution of the program can be resumed. This ability to pause the forward or backward execution at any point and the ability to switch the direction of execution is the generalized challenge of reversible computing. Specific variants and specializations of this general reversible computing problem arise in different contexts. Historically, low power computing has been a major motivating factor behind the development of reversible computing. Over time, additional areas have emerged in which reversible computing has found applications.

2.2   Energy-Optimal Computing

When contemplating ways to reduce the energy consumed to accomplish a certain computation, the thought process naturally leads to the determination of the fundamental lower bound for the same. In this process, it was discovered that the theoretical lower bound on energy is directly related to the reversibility of computation [Landauer, 1961]. More precisely, there are ways in which the original (forward) computation could be relaxed into a more generalized setting of *reversible* computing, (these ideas are elaborated more later in Chapter 4 and Chapter 5). Arguments based on thermodynamics helped keep the treatment beyond any specific hardware technologies [Bennett, 2003]. In an ideal setting, it was shown that no energy need be fundamentally lost in performing any given computation [Bennett, 1973, 1982]. As a corollary, this implies that, in theory, arbitrarily large fractions of the energy consumed for a computation should be recoverable for useful work again, although there would be an underlying trade-off between the efficiency of such recovery and the speed of computation (i.e., computation time approaches infinity as the recoverable fraction approaches unity). Based on the thermodynamics of computation, various hardware technologies have been proposed, designed, and explored, and some have even been implemented as proofs of concept [Ren and Semenov, 2011].

The theory about the minimality of energy spent for computation is built over the most basic bit operation, namely *bit erasure* (e.g., resetting a bit to 0 independent of whether its current bit value is 0 or 1 [Bub, 2001]). First, it is argued that bit erasure is the only operation that fundamentally consumes energy in any computation. An equivalent way of saying the same is the following: any computation can be carried out in such a way that only the bit erasures performed as part of that computation consume irrecoverable energy, while the energy used in all other bit operations is fully recoverable and reusable.

Given that the bit erasures form the sole cause of irrecoverable energy consumption, the important question that follows is whether every computation can be performed without bit erasures. This question was answered in the positive by introducing reversible computing. A notion of the *reverse* of any computation is introduced, which is then used in a higher-level algorithm to accomplish any computation without bit erasures (see Chapter 4 and Chapter 6). In this way, the reduction and recycling of energy is one of the fundamental applications of reversible computing, holding far-reaching potential in the future of computing.

2.3 Parallel Computing and Synchronization

Reversible computing finds use in addressing some of the challenges in large-scale parallel computing such as reduction of synchronization overheads and increasing the concurrency dynamically. As of this writing, parallel computers with millions of processing units is a reality [TOP500.org, 2013], and the scale is only envisioned to increase in the future.

Synchronization is that critical aspect of parallel computation which ensures that the overall execution obeys and incorporates all the inter-processor data dependencies of the application [Grama et al., 2003, Raynal, 2012]. A runtime component of parallel computing is needed in all applications to realize proper synchronization, and this runtime synchronization costs the application some wasted processing time (also called *blocked time*) at some processors during different periods of execution. Some of this synchronization is fundamental in nature, in the sense that the inter-processor data dependencies impose a theoretical total parallel execution time (given by critical path analysis). However, this lower bound is often difficult to achieve in practice, and the programs are written with very conservative order of execution. For example, global barriers are invoked in the program to ensure all processors reach known points in the program before proceeding further. The cost for each such synchronization operation becomes significantly large as the number of processors increases, with some applications spending more than 50% of their time in synchronization versus actual, useful computation.

Note that any synchronization operation is pure overhead, in the sense that it only adds a component of execution time (and energy) that is simply absent in sequential computation. The synchronization problem is an important issue at the highest scales of hardware, despite some advances in underlying network hardware technologies to speed up the synchronization operations themselves.

2.3.1 Asynchronous Computing

The synchronization problem can be solved by finding ways to increase concurrency across processors such that there is increasingly more useful computation performed per synchronization operation. One approach is to incrementally reorganize operations to overlap some computation with communication or synchronization operations. The more general method is to use a relaxation to a generalized asynchronous reversible execution, allowing the application to uncover maximum concurrency at runtime.

■ **Computation–communication overlap** One specialized way to increase concurrency is by using non-blocking synchronization operations, by which processors perform some local operation while a synchronization operation is in progress. For example, non-blocking collective operations such as global barriers or global reductions may be initiated and completed asynchronously. However, the amount of increased concurrency uncovered by this approach is limited to the amount of local computation that can be performed during synchronization.

■ **Generalized asynchronous reversible execution** A generalized way to increase concurrency is to relax the parallel execution model to one in which all processors can execute the application both forward and backward, and let processors execute local computation asynchronously with respect to each other. Communication and synchronization proceed in the background, and any violations of data dependency order are detected at runtime and corrected by relying on reversible execution. Executing backward, the processor is rolled back from its current point of execution to the point of dependency violation, and, after incorporating the correction (e.g., new data arriving from another processor), forward execution is restarted from the corrected point.

In the specialized communication–computation overlap approach, execution follows conventional computation, that is, it is still fundamentally forward execution-based, with no tolerance for any data dependency violation. In the generalized asynchronous reversible execution, the view of execution is fundamentally different, that is, execution is assumed to be reversible from the outset, with a runtime component (supervisor, coordinator, controller, or engine) orchestrating the global execution such that the overall execution eventually provides the same answers as one that incorporated all inter-processor data dependencies. Without reversibility, it is not possible to relax execution to uncover any more concurrency than the most conservative one that avoids even *transient* violations of dependencies.

The best demonstrations of the potential for the reversible execution to reduce synchronization overheads in large-scale parallel computing has been in the area of parallel discrete event simulation (PDES). The Time Warp algorithm [Jefferson, 1985] for PDES has been gainfully employed in multiple large-scale applications (e.g., epidemiological outbreak simulations) using rollback-based synchronization implemented on supercomputers. The algorithm is built on relaxation to a reversible execution, in which local events at a processor are processed optimistically even while lower bound guarantees on the global virtual time are being computed across all processors. Also, theoretical analyses have been performed to show the gains with rollback dynamics in the presence of primary and secondary rollbacks in Time Warp [Akyildiz et al., 1992].

2.3.2   Supercriticality

Critical Path (CP) analysis of a parallel application is the determination of the inherently sequential path(s) of data dependencies from the start to end of the computation [Yang and Miller, 1988, Hollingsworth, 1998]. Clearly, the longest path from start to end is the most critical one. It is well-known that the critical path(s) in any application determine(s) a fundamental lower bound on the parallel application's total computation time. In particular, CP imposes an absolute upper bound on the achievable parallel efficiency or speedup. However, an important aspect about this CP-bounded parallel efficiency is that it rests on the fundamental assumption of conventional forward-only (irreversible) computation. If the assumption of irreversibility is relaxed, and the lower bound is examined with a new assumption of reversible execution, it is in fact possible to sometimes exceed the theoretical CP-based lower bound on computation time (and, consequently, upper bound on parallel efficiency and speedup).

To illustrate, suppose computations appear as a sequence of updates, of the form

$$U_{ij} : x_i \leftarrow f_j(\mathcal{R}_{ij}),$$

where x_i is a variable being updated, $\mathcal{R}_{ij}$ is the set of all variables that x_i depends on for update j, and f_j is a computable expression on that set of variables. In normal execution, the exact values of all variables in $\mathcal{R}_{ij}$ should be available before the update can be performed. In this model, the chain of these dependencies gives the CP-bounded computation time.

Now, consider the new computation model in which the computation is relaxed to be reversible, i.e., f_j^{-1} exists and is available for every j. In this reversible model, suppose U_{ij} is computed without waiting for the most recent values of $\mathcal{R}_{ij}$. After this out-of-order update, when new values R_{ij}^* for $\mathcal{R}_{ij}$ are available, the correct new value for x_i must be recomputed. This is accomplished by first invoking f_j^{-1}, then incorporating R_{ij}^*, and finally recomputing U_{ij}. This reversal and recomputation can reduce synchronization effects via reversible execution, yet it does not defeat the fundamental critical path-based lower bound on computation time.

To improve the execution beyond the CP limit, consider the following variation: when new values R_{ij}^* of $\mathcal{R}_{ij}$ are received, a temporary value $x_i^* = f_j(R_{ij}^*)$ is first computed and compared with the current value of x_i that was computed out of order. If they are equal, that is, if $x_i = x_i^*$, then there is no need to overwrite/update x_i because it already has the correct value. There are two extremely important implications of this obviation of update to x_i:

■ No succeeding nodes in the dependency graph will need updates, that is, no U_{ik}, $j < k$, needs to be recomputed (in case they have already been computed).

■ Due to computation of U_{ij} ahead of its position in the dependency graph, its computation is overlapped with other preceding computations U_{ik} for some $k < j$, thereby reducing the overall completion time of the graph.

This phenomenon of the potential to defeat the lower bound dictated by the critical path is called supercriticality [Jefferson and Reiher, 1991]. It not only preserves the correctness of the final answer, but, more importantly, defeats the lower bound on computation time if U_{ij} is in the critical path of the over-all computation Note that supercriticality is only possible due to reversible execution. In the example, if x_i is not equal to x_i^*, then f^{-1} would have to be invoked to correct the execution to conform to the dependencies. Thus, reversible computing can be used to sometimes provide supercritical paral-lel computation that exceeds the parallel efficiency and speedup otherwise prescribed by the critical path.

2.3.3 Performance Effects

When reversibility is introduced to relieve synchronization costs, the program must be instrumented to enable forward as well as backward modes of compu-tation. Two broad approaches are (1) saving copies of variables to a memory trace before they are modified in the forward mode and copying back from the trace in reverse mode, or (2) avoid saving values of variables in forward mode, but instead invoke inverses of individual operations in reverse mode. In hardware technologies that have high memory latencies (i.e., time taken to read from or write to memory locations) compared to speed of compu-tation (such as computing arithmetic operations), it is vastly more efficient to perform inverse computations rather than store and retrieve values from memory.

The so-called "memory wall" being faced in the 2000s-2010s exhibits this situation in which the increasing speeds of central processing units (CPUs) make memory operations relatively very expensive. This is because of the in-creasing ratio of CPU clock rates compared to those of memory units, and also because of multi-core technologies that result in an increase in the number of CPUs accessing the same number of memory units. This widening gap be-tween the speeds of arithmetic/logic operations and memory storage/retrieval operations can be bridged via reversible computing.

Operationally, the performance differences manifest themselves in terms of memory behavior such as caching effects at all levels (L1, L2, L3), and Translation Look-aside Buffer (TLB) effects. Memory-based (checkpointing) reversal suffers from the following drawbacks: (1) the total amount of memory needed for the reversible program is larger than that without reversible execu-tion; (2) the time cost of copying values before being modified in the forward mode adds to the forward execution overhead, potentially slowing down for-ward execution; and (3) memory subsystem behavior (cache and TLB) can be negatively impacted due to decrease in locality and due to memory accesses spilling over cache size limits. Computation-based reversal can relieve these drawbacks because (1) little additional memory is needed for reversal of arith-metic operations; (2) the forward code is largely unaffected, thus retaining the original forward execution speed; and (3) memory subsystem effects are not significantly altered from the original forward mode.

2.4 Processor Architectures

In processor architecture technology, the need for reversible execution arises in two distinct ways: (1) speculative execution, and (2) very large instruction word. In both cases, reversible execution is used to correctly uncover dynamic parallelism from an otherwise sequentially specified computation.

2.4.1 Speculative Execution

Consider a single sequence of instructions $\mathcal{S} = < \ldots, I_i, \ldots >$ being fetched, decoded, and executed by a processor. In general, instruction I_i must be executed only after all instructions $I_j, j < i$ are completed, because the set of variables I_i depends on may be modified by those prior instructions. However, such dependency may not always be actually present in a small window of the sequence, and some of the instructions may be safely processed concurrently without waiting for the others to finish. For example, $I_1 = \boxed{a \leftarrow b \times c}$ and $I_2 = \boxed{d \leftarrow e + f}$ can be processed concurrently even if specified sequentially because there is no intersection of variables being read or modified. More complex instances involve implied dependencies such as from array operations, dereferences, and register conflicts. The problem is that the correct execution order cannot be determined statically, a priori. In general, the set of variables being modified by an instruction is not necessarily known until it is actually decoded and executed (e.g., when indirect references such as array indices or pointers are used to refer to the variable being read or modified). Thus, there is a conflict between the possibility of executing a few instructions concurrently to increase the overall processing speed and the possibility of the concurrently executed instructions incorrectly affecting each other that might result in wrong results.

A way to dynamically exploit the potential concurrency is via *speculative execution* in which a later instruction is issued even before the earlier instructions are completed [Dubois et al., 2012]. After the speculatively executed instruction is executed, conflict detection is performed to see whether there was an intersection in the set of affected variables (i.e., a violation in read/write dependencies). If a conflict is detected, the results of the speculatively executed instruction are quashed and that instruction is restarted from the beginning. Alternatively, fix-up code may be invoked to repair the incorrectly (speculatively) computed results, instead of discarding everything and restarting the speculatively executed instructions from scratch. A fair amount of processor infrastructure and compiler support is usually needed to accomplish speculative execution. For example, at the processor-level, instruction execution must be relaxed so that modifications to registers are held in temporary (shadow) registers for every speculatively executed instruction, and the changes are committed from the temporary to actual registers only when conflict resolution succeeds. Reverse computation could be used to avoid these temporary registers by invoking an inverse instruction to restore the affected register value if the speculative instruction was found to be conflicting. Speculative execution is widely used in many modern processors to increase the processor instruction throughput.

2.4.2   Very Large Instruction Word

In a manner complementary to speculative execution, suppose each instruction I_i is in fact a vector of sub-instructions, $I_i = [s_{i1}, \ldots, s_{ij}, \ldots, s_{in}]$, where the sub-instructions s_{ij} can all be processed concurrently to constitute the execution of I_i. Because the number of sub-instructions can be large, the width of each instruction becomes large, giving the name *Very Large Instruction Word* (VLIW) [Fisher, 1983, Dubois et al., 2012]. Determination of which sub-instructions are possible to execute as an aggregate instruction is accomplished within the compiler that generates the VLIW code. While the *processor* is responsible for determining the concurrency and reversal of instructions in speculative execution systems, the *compiler* is responsible to determine the concurrency and reversal of sub-instructions in VLIW systems. The VLIW compiler attempts to generate instructions that have a dense packing of sub-instructions; however, it may have to reorder sub-instructions in generating such dense packings. The compiler is then responsible for generating compensation code that reverses the effects of incorrectly (speculatively) ordered sub-instructions if they result in data conflicts at runtime. The most well-known packing method is the Trace Scheduling approach [Fisher, 1981, 1983] in which the path taken by the code in an execution is used to generate the basic instruction sequence (independent of branch conditions), and reversal or compensatory code is added to this sequence to recover from deviations from the assumed path.

2.4.3   Anti-Memoization

A form of reversible computing can be employed in recovering temporary values that are usually pushed from registers to main memory when a register conflict occurs (due to register file size limitation or due to named registers being reused and overwritten across operations). If a register value is potentially about to be lost, it is usually pushed to memory and later loaded back from memory when it is again needed/used. Because memory operations are orders of magnitude slower than register speeds, it is desirable to keep operations confined to register accesses as much as possible. Thus, when the intermediate value that is lost in a register (due to being overwritten) is again needed, it is sometimes possible to either recompute or reverse compute earlier operations to recover the lost register value, instead of relying on storage to/retrieval from main memory.

A generalization of such memory *versus* computation trade-off is the concept of memoization and anti-memoization. The method of storing the value of an intermediate computation in memory for reuse in later computation is called memoization [Hoffmann, 1992] (note that the term *memoization* is different from *memorization*). Such an optimization that trades off memory for computation arises either from explicit programmer intent or from automation techniques such as code generation by compilers. While memoization works best when memory is cheaper than recomputation, it can in fact degrade performance compared to recomputation when memory access cost is much higher than recomputation cost. However, if the code has already been written using memoization, the memoized values must be recovered using recomputation. This can be done via reverse computing to the most recent position at which the variable was last stored to memory, and then recomputing the expression that led to the stored value. We call this approach *anti-memoization*, which is the process of undoing memoization. An instance of this high-level approach has been applied at the level of register value recovery (called register rematerialization) [Bahi and Eisenbeis, 2011, 2012] to improve the overall application performance.

2.5 Debugging

Reversibility of execution is very useful in the process of debugging programs in an efficient and convenient fashion. When running a program, if an unexpected condition or undesirable results occur, the program's execution needs to be retraced to find the precise point where the deviation from desired behavior actually originated. In order to be able to step backward, the program state needs to be saved before every forward operation. However, the amount of saved state can become extremely large because the computer executes millions of instructions per second; in fact, the accumulated state grows so quickly that it may be infeasible to store the program trace for long program execution length. The trace for only a small window of execution may be stored. Yet, the distance between the point of the bug's manifestation and the original source location of the bug may be large, making it infeasible to step backward to the correct origin of the bug via trace traversal. This is where reversible computing can be applied—instead of saving/restoring the values of the variables to/from memory, the inverses of the instructions can be executed backward to traverse the program in reverse from the bug manifestation point until the bug's source is determined.

The difficulty in debugging is especially pronounced in (1) assembly-level debugging, due to very large sizes of traces; and (2) parallel computing. Reversible computing is either a significantly more efficient method or the only feasible method in these cases.

In debugging programs at the level of its assembly language, since the number of assembly instructions or machine code is extremely large, the trace sizes needed to enable backward traversal in assembly code grows extremely quickly. Reversible computing is the only feasible method to enable bi-directional movement in assembly instruction streams for efficient debugging. Without reversible computing, assembly-level debugging either slows down forward execution tremendously (due to the introduction of the high cost of trace generation operations before every assembly instruction) or is infeasible because the trace does not fit in the available computer memory.

The problem of debugging has been described and various methods surveyed extensively in the context of high-level programs such as text editors and user interface systems [Teitelman, 1975, 1984, Archer et al., 1984, Leeman, 1986]. General bi-directional movement for debugging in general has been studied [Boothe, 2000], especially taking care of logging all the relevant states including system call data to provide determinism in repeated bi-directional movement across already-executed instructions. Assembly-level debugging via reverse execution was studied and optimizations proposed to significantly reduce the amount of memory trace needed for backward traversal [Akgul and Mooney III, 2004, Lee, 2007]. While reversible execution has been successfully applied in parallel computing applications (e.g., [Carothers et al., 1999]), full-scale application of reversible debugging in large parallel systems is a relatively open item of research.

2.6   Source Code Control Systems

Source code control systems provide a world view in which modifications to a set of objects can be tracked, traversed, and manipulated along different logical timelines. The timelines are formed by sequences of individual or grouped modifications to objects. In general, the timelines are related to each other in the form of directed acyclic graphs (DAGs). Most often, each object is a named file in a file system. Modifications to the file are characterized in terms of edits or changes to the individual lines, assuming that the file is a text file. Popular source code control systems such as `git` [Loeliger and McCullough, 2012, Somasundaram, 2013] and `mercurial` [O'Sullivan, 2009], as well as older systems such as `svn` [Collins-Sussman et al., 2009] and `cvs` [Thomas and Hunt, 2003], provide reversible sequences of edits to sets of files. They also provide ways to identify and extract, via user-specified or system-generated naming, individual paths along the DAG of timelines.

Overall, the systems provide a way to view the evolution of the object values as a reversible computation that can be traversed forward or backward and also merge different timelines. When each object is viewed as a program variable holding data, changes to the values of the variables can be made in a reversible fashion to save and restore the data values. Thus, although the granularity of the objects is different from the variables in a conventional computer program (files often have much larger bit lengths than typical program variables), the notion of reversibility in source code control systems is analogous to that of a program's state evolution. Concepts common to both include the *undo* and *commit* operations, while the *branching* and *merging* are operations that are somewhat specific to source code control systems.

2.7   Fault Detection

In fault detection, the idea is to utilize reversible computing to periodically verify if the forward computation was performed correctly, as follows. Given a code fragment P, after it has been executed forward as $F(P)$, it is to be ascertained if there had been a faulty execution of any portion of P (for example, values of some variables may become incorrect due to low-probability errors in the memory subsystem that flip one or more bits randomly due to electrical faults). If the code is executed backward as $R(F(P))$, then the initial values of the variables prior to execution of $F(P)$ will not match the values restored by $R(F(P))$; such a mismatch can be used to signal an error condition. This method of detection relies on two important, reasonable assumptions:

1. The type of errors encountered in the forward path are rare events, and hence the backward path is not susceptible to the same errors as well. Because $F(P)$ executed under errors and $R(P)$ did not, their net results can be expected to differ from each other.

2. In the rare event that the backward path also encounters errors, we can reasonably assume that the forward errors and backward errors do not cancel each other.

Reversible computing can be used for fault detection and, in some cases, fault correction. Reverse execution can be used in trapping errors in either forward code or the reverse code, or even in the implementation of the reverse compiler. This is achieved by checking the following simple *necessary* correctness condition at runtime [Bishop, 1997], which, when applied on every local variable and global variable, is useful in detecting errors in the code or the compiler.

> Suppose a variable `var` is initialized to `expression` in the forward execution. At the end of the reverse execution, the final value of the variable `var` must equal `expression`.

Another simple correctness condition is the following, which is a generalization of the end-of-tape condition described in [Bishop, 1997]:

Suppose the bit/byte tape is at position P before the forward execution of a function f, and later the reverse of the function is executed. Then, by the end of the reverse execution, the tape must be rewound to the same position P.

2.8 Fault Tolerance

Fault tolerant computation in a parallel or distributed system is the ability to gracefully continue execution of an application despite transient faults or failures of system components at runtime. Fault tolerance is an extremely difficult capability to achieve in parallel systems, particularly when the number of components in the system is very large. Simplistic schemes rely on periodically saving the entire application state to persistent storage and restoring this state at all processors for recovering from failures. However, such schemes are woefully non-scalable and break down with large numbers of processors. More scalable solutions do not rely on global checkpoint/restart views, but use in-memory solutions. Among them, rollback-based recovery is an important algorithmic core underlying scalable parallel computing, appearing in the form of system support, middleware, or applications. For example, efficient rollback-based fault tolerance approaches (e.g., [Manivannan and Singhal, 1996, Kim et al., 1996], among many others) rely critically on the ability of processors to revert their state back to a point in the past.

Thus, processors need the ability to go back to a previous point in execution dynamically on demand, when they are informed of a *fault*. The definition of a fault varies with application. Often, a fault is the detection of a failure of a processor. In other software-level rollback schemes, a fault is the detection of a violation of application-specific event order for correctness. For example, in large-scale Time Warp [Jefferson, 1985, Perumalla, 2007], a *primary rollback* results when an event is received with timestamp smaller than the current virtual time of the receiving processor. When previously sent messages are taken back as a result of primary rollback, further rollbacks, called *secondary rollbacks*, are transitively propagated to other processors.

Figure 2.1 shows the schematic for this general setting [Perumalla and Park, 2013]. When a processor P_f encounters a fault, it restarts from the most recently saved checkpoint LC_f and informs all other processors $\{P_r\}$ to roll back to the point corresponding to the program state of LC_f. Every rolling back processor P_r can invoke reverse code to recover the state corresponding to LC_f.

Note that LC_f and LC_r are in general different because, for maximum efficiency, each processor is allowed to asynchronously and *infrequently* initiate a checkpoint of its own state to persistent storage. Note also that every rolling

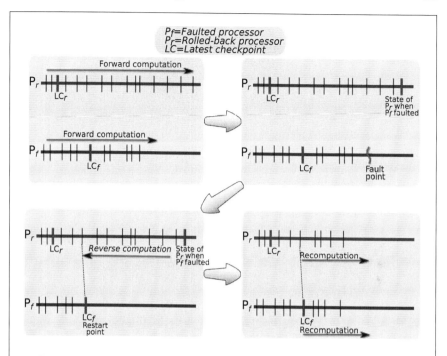

FIGURE 2.1: Schematic of asynchronous recovery sequence using reverse computation-based rollback.

Each thin vertical line denotes an update to the local state of a processor. Each thick vertical line marks a full checkpoint of the processor state to persistent storage. Note that there are very many processors indicated by P_r that are affected due to the fault at P_f. All of them need to be rolled back, although, for simplicity, only one processor is shown in the illustration.

back (non-faulted) processor P_r need only use reverse computation, but does *not* have to access its own checkpoint LC_r. The checkpoint is only used by a processor if and only if it is the faulted processor. This particular aspect *dramatically* relieves congestion in the system in terms of lowered pressure on the memory bus and on the file system.

In Figure 2.1, P_r's corresponding state of LC_f is joined by the dashed line between the processors. The faulted processor P_f restarts from LC_f, because the state from LC_f to the fault point is assumed to be lost or unavailable due to failure.

2.9   Database Transactions

The concept of reversal of operations is fundamental to the backbone of databases, namely, database transactions. Databases provide the key properties of atomicity, consistency, isolation, and durability (ACID properties) for any groups of operations called transactions [Date, 2003]. Applications can be written correctly and conveniently using these properties. Database systems internally provide sophisticated and elaborate implementations to provide support for *transactions* [Berstein and Newcomer, 2009], fundamentally relying on the concepts of reversal and replay of indivisible groups of operations. Operations are logged into persistent storage, and complex algorithms ensure that the state of the logs and the state of the data stored in the database are always consistent with each other. Reversibility of a transaction is key to correct operation because a transaction may be aborted at any time (either intentionally by the user due to change of mind, or unintentionally by the user such as from loss of network connection, or automatically by the system due to faults such as power failures). When a transaction is aborted mid-way, all operations performed as part of the transaction must be undone to preserve the critical ACID properties.

In a common example often used to illustrate the reversal of operations for transactions, two transactions T_1 and T_2 operate on a database of two bank accounts A_1 and A_2. Transaction T_1 attempts to transfer x dollars from A_1 to A_2, while T_2 attempts to transfer y dollars from A_2 to A_1. To transfer, T_1 first debits A_1 by x dollars and then credits A_2 by the same amount. Similarly, T_2 first debits A_2 by y dollars and credits A_1 by the same amount. Thus, $T_1 = \boxed{A_1 \leftarrow A_1 - x; A_2 \leftarrow A_2 + x}$ and $T_2 = \boxed{A_2 \leftarrow A_2 - y; A_1 \leftarrow A_2 + y}$. If the database state before either transaction is $\boxed{S}$, then the transaction system ensures that the final system state after the transactions is only one among $\boxed{S \mapsto T_1}$, $\boxed{S \mapsto T_2}$, $\boxed{S \mapsto T_1 \mapsto T_2}$, or $\boxed{S \mapsto T_2 \mapsto T_1}$, where $\boxed{P \mapsto Q}$ denotes the state obtained by application of transaction Q on state P. In other words, it ensures that at most one transaction succeeds, or if both succeed, the state is exactly *as though* one transaction is entirely preceded by the other (i.e., not interleaved). To achieve these semantics, reversal of operations is employed if and when any transaction is aborted before it is executed to completion (i.e., "committed"). For example, if T_1 is aborted after its first step $\boxed{A_1 \leftarrow A_1 - x}$, this partially executed transaction can be undone by executing the inverse $\boxed{A_1 \leftarrow A_1 + x}$. Similarly, if T_1 completes both steps, but somehow fails before it is committed, both steps are to be undone, in reverse order; That is, by executing the inverse $\boxed{A_2 \leftarrow A_2 - x; A_1 \leftarrow A_1 + x}$. While this simple example illustrates the reversal of aborted transactions in databases, in practice the database system infrastructure is much more elaborate and complex to support very fast operation of a large number of transactions containing a richer set of operations. Different reversal technologies are employed to roll back transactions, the superset of which is in the realm of reversible computing.

2.10 Quantum Computing

Reversible computing is an inherent feature of Quantum Computing [Bennett et al., 1997, Rieffel and Polak, 2011]. In Quantum Computing, computation is a sequence of *unitary* operation of the computer state. Because every unitary matrix is reversible by definition, the entire sequence is inherently reversible.

2.11 Additional Applications

Reversible computing finds use in different forms and to varying degrees in several other applications. The adjoint methods in Automatic Differentiation (AD) can exploit reversible computation in the so-called reverse mode of evaluation [Griewank and Walther, 2008]. User-friendly graphical interface-based applications commonly provide interfaces to allow users to perform certain operations that can be undone on demand, allowing the user to explore different operations. Sports scoreboard maintenance and recording systems are built using reversible computing principles to enable automated generation of inverse actions for normal actions, to deal with inherently error-prone processes in real-time scoring [Briggs, 1987]. Computer file systems, such as the Apple Macintosh Operating System's Time Machine® functionality, provide a reversible view of all changes to the file system contents.

Interesting Contexts

Software for reversible execution found an interesting application in the 1980s when J. Briggs reported a way to use reverse code generation and reversible execution to undo incorrect updates to the scoreboard in cricket matches [Briggs, 1987]. Reversal code was automatically generated with the objective of minimization of the state information to be stored to enable reversal.

In the game of cricket [Knight et al., 2007], as in other sports, mistakes and corrections inevitably occur in recording and publishing the scores even as the game is in progress. Errors can appear in two ways: (1) the scoreboard operators may commit errors of omission or commission in entering events into the computing system, or (2) the game itself may experience reversals of decisions and other sport-specific updates, such as umpire's corrections. In each of these types, there are many occasions where updates are rolled back and corrected, naturally warranting reversible execution in the scoring program software.

In an unrelated context, reversibility appears in musical compositions in the concepts of the *mirror canon* and the *crab canon* (also called *cancrizans*) in which the musical notes are mirror images of themselves (or palindromic) [Hugo, 1904]. A popular reference to crab canons is the collection by J. S. Bach titled "The Musical Offering" [Bach, 1747]. The notes of a crab canon written over a Möbius strip [Pickover, 2007] can be played back and forth ad infinitum.

Chapter 3

Reversible Computing Spectrum

3.1 Spectrum

The spectrum of reversible computing is wide, spanning from programming languages that are closest to the applications, down to hardware circuitry that ultimately realizes the computation in the form of modeled physical processes. The components in the spectrum are illustrated in Figure 3.1, which shows them in traditional (irreversible) computing in correspondence with those in reversible computing. There are also possibilities to transition from the irreversible column to the reversible column, and vice versa, at different levels of the spectrum. These transition options give rise to possibilities of realizing reversible computation over irreversible computation, and vice versa, in the intermediate points of the spectrum.

3.1.1 Components

At the top of the spectrum is the set of programming languages: conventional languages such as **C** and **C**$^{++}$ are irreversible in general, while specially defined languages such as **Janus** are entirely reversible by design. A subset of programs written in irreversible languages can be reversible (if, for example, they only utilize reversible operations). Theoretically, those programs can be translated into reversible languages. Thus, we have the set of irreversible programs and the set of reversible programs; the set of reversible programs can be written in a specially designed reversible language or in a subset of an irreversible language.

The compilation of programs itself could be qualified as reversible or irreversible. When the compiler is viewed as a program that accepts source code as input and generates object code as output, the execution of the compiler program could itself be reversible or irreversible. A reversible compiler accepts source code as input, generates object code, makes a copy of the object code to the output, and uncomputes back, leaving only the object code as output. The best example of a reversible compiler is the **Janus** interpreter [Yokoyama and Glück, 2007] that itself is written in the reversible **Janus** (hence, also called a self-interpreter). Almost all traditional compilers, such as the GNU C Compiler, are irreversible compilers. Note that some irreversible compilers can generate object code for reversible programs. For example, source-to-source translators such as the Reverse C Compiler (RCC) execute irreversibly but generate reversible code.

The target of the compilers for object code can be conventional irreversible instruction set architectures (ISA), such as the Intel X86 ISA, or they can be reversible instruction sets, such as the Pendulum Instruction Set Architecture (PISA) [Vieri, 1995, Vieri et al., 1998, Vieri, 1999, Frank, 1999]. Reversible ISA can be natively executed reversibly on reversible hardware, but they can also be executed irreversibly on irreversible hardware. Similarly, irreversible ISA can be emulated reversibly on reversible hardware, and, of course, irreversibly executed on conventional irreversible hardware. This is the last layer at which transitions between irreversible execution and reversible execution spectrum can happen. Below this layer, the nature of computation with respect to reversibility cannot be changed.

The hardware technologies for computing come in both flavors: (1) conventional irreversible gates, or (2) reversible gates providing assured reversibility (potentially with consequent thermodynamic implications such as low energy consumption). The NAND and NOR universal gates are examples of irreversible gates, while CNOT and CCNOT family of gates are well-known examples of reversible gates.

Using the basic gates, large circuits are built for general-purpose and complex computations. These circuits are necessarily distinguished as reversible or irreversible, depending on the specific types of gates used. Large circuits are synthesized from the basic gates as building blocks, with different synthesis technologies that are different for reversible and irreversible gates.

3.1.2   Common Cases

For reversible computing, the common case has programs written in a reversible programming language, compiled with an irreversible compiler on an irreversible computer, with the resulting program executed on a reversible computer supporting a reversible instruction set. In another common case, a reversible program is generated by an irreversible compiler executed on an irreversible computer, and the resulting reversible program is used in an otherwise irreversible, larger program executed on an irreversible computer. A few demonstrations have included actual reversible execution all the way from the programming language level down to the level of reversible circuits, with the exception of the compilation process that is typically performed with an irreversible compiler on an irreversible computer.

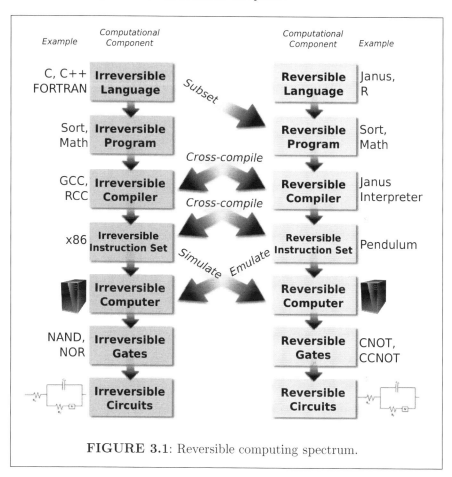

FIGURE 3.1: Reversible computing spectrum.

3.2 Partial Reversibility

Between the two extremes of irreversible computation and fully reversible computation, one can contemplate an intermediate notion of *partial reversibility*. In partial reversibility, two options exist: (1) a small loss of information is allowed to occur during execution, and (2) a small amount of memory trace is allowed to accumulate during execution.

In the former, the execution is mostly reversible except for the loss of a relatively small number of bits compared to the number of bits comprising the entire state of the system. For example, in a simulation of molecular dynamics of millions of particles, the number of bits comprising the state of the whole system of particles may be extremely large. However, perfectly reversible execution may not be possible because of the complexity of modeling certain collision operators such as a multi-particle collision involving three or more particles simultaneously. Although irreversible processing of such collisions results in a loss of information, the error may be acceptable because such collisions are rare and may not adversely affect the overall dynamics of interest.

In the latter, the relatively small amount of information that is being lost in the rare dynamic conditions may be recorded in memory to be able to reverse them. Although theoretically the trace length is proportional to the execution length, the size of the trace may be extremely small in practice and hence negligible [Perumalla and Protopopescu, 2013].

Similar considerations can be applied as well to the synthesis of circuits for low power computing. In reversible hardware designs that reduce energy usage in exchange for a lower speed of computation, partial (ir)reversibility in the form of (ir)recoverable energy is possible, which may provide a better trade-off between energy savings and computation speed or memory size [Li and Vitanyi, 1996, Buhrman et al., 2001].

3.3   Unit of Reversibility

The notion of reversibility (at least in relation to the inanimate world) is rather relative. Indeed, many levels of reversibility, and thereby irreversibility, can be defined without contradiction or inconsistency across the levels. The levels are defined by the level of abstraction or amount of detail one chooses to use to describe a system and view it as such.

Because computing involves transformation of an input vector to an output vector, one could easily propose a trivial form of reversibility in which a copy of the input vector is saved before computation, and restored when computation is to be undone. There are two assumptions underlying this view: (1) the computation overwrites the input vector with the output vector, and (2) an unmodified copy of the input vector is unavailable elsewhere.

3.3.1   Reversing a Child's Play

Consider a child playing with toys in his room. Before he starts playing, the room is neatly organized, with every item of the room in its prescribed place. After, say, an hour of playing, the room gets quite disorganized, with toys scattered in various directions. Suppose one were to pose the question of reversibility of the child's play. The unit of reversibility becomes important here. If the unit is the whole hour's play, then it is easy to achieve reversibility: simply reset all the items to their original, prescribed places—we know that they started that way before the child started playing, and so, by restoring all the toys to their original places, we have reversed the child's entire playing activity. On the other hand, if the unit of reversibility is any finer than the whole hour of play, then the reversibility problem gets significantly harder. For example, if we must be prepared to restore the room to any intermediate position x minutes into play, where x is specified only after all the play is done, then the activity needs to be tracked at a fine time resolution throughout the play. Issues of interpolation also arise if camera-based snapshots are not sufficiently frequent, and so on.

3.3.2 Reversing the Movement of Library Books

Another example that illustrates how the desired unit of reversibility determines the difficulty or ease of reversal is the restoration of books to their numerically sorted order in a library. After the movement of books by people within a library over the course of a day, if the unit of reversal is the entire day, then reversal is trivially achieved simply by picking up every book that is not in its correct place and placing it back in its right place within its designated shelf. On the other hand, if the reversal problem is generalized to that of restoring the state of the library any x minutes into the day, then the problem becomes much more difficult. The movement of the books must be recorded sufficiently often to remember the location of every book at every instant of time in the day, in anticipation of the requirement to restore the state to any point in time. The information to be remembered includes the identity of every book that is being moved, its current location, its new location, and the time of the movement. Thus, what required zero memory for reversal of an entire day's movement now requires a trace of all movement of all books within the day.

3.3.3 Reversing Different Units of Computation

Analogous to the preceding examples, the problem of reversal varies with the unit of computation that needs to be reversed in reversible computing. The cost of reversibility in computing depends on the specific usage context and the specific unit of reversal. For example, any user process (e.g., UNIX process) can be reversed simply by resetting the memory and restarting from the beginning. However, the difficulty of reversal increases if the process may be paused during execution to be rolled back to an arbitrary point in its past. To reverse it to an intermediate point in the past, the computation must be tracked and the state of the reversal point must be accurately recreated upon reversal. Thus, the unit of reversal can range all the way from the transformations at the bit-level to the level of processes or even higher (e.g., rebooting the computer). The objective and the unit of computation together determine the reversal semantics.

Part II

Theory

Chapter 4

Systems and Principles

4.1   Logical Computations and Physical Processes

In pursuit of scientific inquiry into the *physics* of the universe, one encounters a phase at which reversibility plays an important role. Quantum mechanics and the Maxwell's Demon are some salient indicators of this intriguing facet of Nature. Equally intriguing, a pursuit of scientific inquiry into the thermo-dynamic limits of *computing* also leads to the role of reversibility in terms of information represented in physical processes. A provocative discussion by Charles Bennett titled "Is Information Physical or Physics Informational?"

highlights, among other things, this uncanny yet critical relation [Bennett, 2005]. Thus, strangely, reversibility appears in the center of both concepts (see Figure 4.1): the informational aspects of phenomena in physics and the energy limits in the physics of computation. Historically, the former was investigated in great depth in relation to physical systems as varied as steam engines and astrophysical bodies. Along the way, principles that transcend almost all technology-specific details evolved in the form of concepts such as entropy and laws of thermodynamics. Much later, the latter appeared when the physical needs of information processing were investigated in the context of energy-efficient computation.

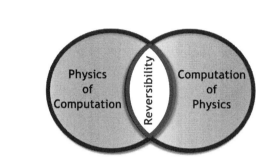

FIGURE 4.1: Reversibility at the intersection of computation and physics.

4.2   System Theoretic View of Computation

A major scientific advance closely tied to reversible computing is in understanding the theoretical limits on the energy needs of computation, independently of any specific computing hardware technologies. The limits are understood by applying the principles of thermodynamics to the physical processes used to realize computation, specifically the energy components that underlie any computation. It is envisioned by many researchers that the theoretical limits are destined to be reached in practice within a decade or two when computing capabilities (e.g., number of bit operations per second per unit volume) are scaled by two to three orders of magnitude relative to today's scale (see, for example, [Frank, 1999, 2005]). Hence, an understanding of the theoretical limits is seen as relevant today in order to better prepare for the impending practical limitations.

4.2.1 A Computation Example

Consider a simple computational problem, namely the computation of the square root of a number. Suppose we have an algorithm $\mathcal{A}$ that accepts a number x as input and computes $y = \sqrt{x}$ as output. Now we ask the question: what is the minimal amount of energy needed (on average, per input) to execute $\mathcal{A}$ on a computer? Alternatively, we may ask what exactly is the nature of energy consumption within the computer (e.g., where does the energy go?).

There are clearly many parameters to take into account in answering the question. For example, there is a wide range of computing technologies that could be used, such as mechanical machines or electrical/electronic devices or even non-traditional technologies such as DNA-based computing [Paun et al., 2006]. The algorithm $\mathcal{A}$ may manifest itself as several logical operations, each of which may consume energy for execution. The condition of the computer's environment (e.g., temperature) may affect the energy needs. And so on. Is it possible to answer such a seemingly open-ended question?

To arrive at an answer, we will first review the components of energy consumption in a computing system, and then examine the fundamental results about the minimal energy needed for any computation.

4.2.2 Basic Components of Computational Energy

Let us examine the elements of energy consumption in a practical computer and make a distinction between the theoretical and engineering-specific components of the energy in actual operation.

Figure 4.2 shows the abstract operation of any computing device, which transforms input bit sequences into output bit sequences based on some set of physical processes that use the following energy components. For example, when algorithm $\mathcal{A}$ is executed, the input bit sequence consists of the list of w-bit integers $x_1, \ldots, x_n$, and the output bit sequence consists of $y_1, \ldots, y_n$, where $y_i = \lfloor \sqrt{x_i} \rfloor$, where $\lfloor x \rfloor$ is the largest integer less than or equal to x. The energy components are as follows:

1. Energy, E_{in}, is input to the computing system to drive the process of computation. For example, this would include the electrical energy to drive the electro-mechanical components, and to run the cooling system that maintains a safe operating temperature for the electro-mechanical computing devices.

2. An irreversibly lost portion of energy, E_{irr}, is predicated as a fundamental loss that is solely a function of the logical transformation function on the input bits and the operating temperature of the device, and independent of all physical characteristics of the device. This is the theoretical

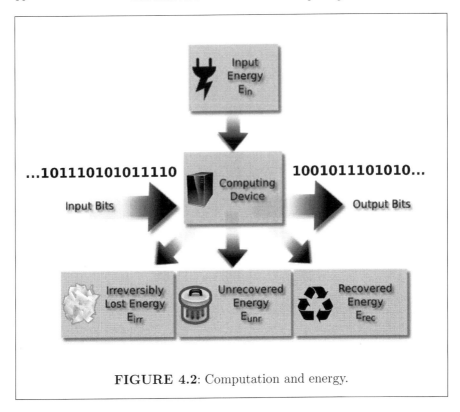

FIGURE 4.2: Computation and energy.

minimum amount of loss that would be necessarily incurred by any computational hardware, independent of efficiencies and engineering details.

3. An unrecovered portion, E_{unr}, of the input energy represents the component that can be traced to engineering limitations, and is subject to change with technologies employed to realize the computation. This portion can be reduced via better design and engineering, and sometimes by adopting different physical processes. At present, a major fraction of energy consumed in modern computers is in the form of unrecovered energy.

4. A recovered portion, E_{rec}, of the input energy that is fed back into the system as part of E_{in}. This energy component is also an artifact of design and engineering. The recovery could be realized by tapping into some of the energy stored in the machine while changing the computing states, and also by tapping into the thermal energy dissipated as heat from the computer. For example, steam could be generated from the dissipated heat for partial conversion back to electrical energy. In many modern installations, the fraction of recoverable/recovered energy is often too small to be useful in feeding back into actual operation of the computing

infrastructure. However, as the efficiency of computation improves, this fraction may become significant.

From conservation of energy, $E_{in} = E_{irr} + E_{unr} + E_{rec}$. In most existing computing systems, $E_{unr} + E_{rec} \gg E_{irr}$ [Frank, 1999]. Also, usually $E_{rec} \ll E_{in}$.

As a side note, an important engineering problem is cooling the computer. Energy lost in the form of E_{unr} not only wastes energy, but also (more importantly) often manifests itself as heat. This heat raises the temperature of the computing system and can make it difficult to cool down the machine sufficiently rapidly when computational hardware density increases, that is, when heat generated per unit volume increases. Thus, even if it is possible to increase the input energy E_{in} to meet the needs of the computation, the unrecovered energy E_{unr} may become so large that it accumulates heat to a level sufficient to melt the computer. This consideration imposes limits on input energy different from the considerations of efficiency and infrastructural capacity.

4.2.3 Dissipated Energy as Theoretical Energy Cost of Computation

To set the lower bound on the theoretical energy cost of any computation (independent of physical process technologies or current design limitations), it is reasonable to assume that appropriate physical processes can be designed and engineered to either store or transform energy from one type to another without losing the ability to control all the energy to do useful work. In other words, all energy in the computing system will remain in a form usable to do computation work, except, potentially, a portion of energy that is postulated as necessarily "lost" due to some fundamental causes arising from the logic of computation. Thus, we may theoretically assume an implementation in which there is no unrecovered energy (making $E_{unr} = 0$) and all recovered energy is recycled into input energy (which may be viewed as $E_{rec} = 0$). The energy component E_{irr} is entirely a property of the computation and *not* that of the physical manifestation of the comptuation.

In this light, the energy cost E_{irr} is called *dissipated energy*, that is, energy on which we lost control to do useful work solely due to the fact that the energy was used as part of computation. Hence, the energy cost of computation is the amount of dissipated energy in accomplishing that computation.

4.2.4 Theoretical Lower Bound on Dissipated Energy

If we turn our attention to the theoretical limits, it is not immediately clear what would reduce E_{irr}. In particular, how low can E_{irr} be made *in principle*? This question was answered in a set of during in the period from the 1950s to the 1980s (see Bennett [1988] for a recount). The development may be viewed as taking three main steps.

1. A hypothesis was initially prevalent that every logical operation or measurement incurred an energy cost [Brillouin, 1956]. For example, John von Neumann opined that every "elementary act of information" performed at an operating temperature T (Kelvin) necessarily dissipates at least $kT \ln 2$ joules, where k is the Boltzmann constant [Neumann, 1966].

2. It was subsequently identified that only a certain type of logic operations results in energy cost because they destroy information, but other logic operations may be performed in principle without dissipating energy. Specifically, only operations of the form $x \leftarrow 0$ or $x \leftarrow 1$ result in energy dissipation, given that the bit value currently held in the location x is not known to be available anywhere else [Landauer, 1961, Bennett, 1973, 1982, 1988, 2003]. Such operations are called *bit erasures*. The energy cost is $kT \ln 2$ joules for every such bit erasure performed by the computation [Landauer, 1961]. This result, known as Landauer's Principle, is the one of the most well-known, direct relations between physical (thermodynamic) systems and logical computation.[1]

3. The remaining issue to settle the question about the energy lower bound was whether there is any computation-specific lower bound on the number of bits erased by a program. This was settled by an even stronger result that showed that *any* computation can be carried out with *zero* bit erasures [Bennett, 1973, 1982].

The developments were based on system-theoretic arguments, which helped overcome the difficulty of technology-specific analyses. Thus, E_{irr} has been theoretically shown to be proportional to the number of bit erasures performed by a computation, independent of the specific computation technology, and that E_{irr} can be made arbitrarily close to zero by avoiding bit erasures. However, the theoretical elimination of dissipated energy is only approached asymptotically by sufficiently slowing the physical processes to avoid energy dissipation. This aspect of "asymptotically zero-energy" computation has earned reversible computing the name of adiabatic computing.

4.2.5   Reversibility for Zero Dissipated Energy

In the preceding development of the lower bound on the energy cost of computation, the notion of reversibility appears in the arguments as follows. To establish the lower bound, it is necessary to choose a computational hardware

---

[1] Nearly four decades after Landauer's Principle was proposed, an experimental verification of the energy dissipation model has been successfully undertaken and reported [Bérut et al., 2012, Drechsler and Wille, 2012].

technology whose physical operation (independent of the logical functions in computation whose energy bounds we are attempting to establish) admits the potential for incurring the least energy cost (ideally, zero). To this end, the availability of conservative processes is assumed for the physical realization of computation, which ensure that no energy need be lost at the physical operation per se, ignoring computational viewpoints for the moment. Also, time-reversible physical operation is assumed for the processes. Several technologies such as ferrites, ferroelectrics, and thin magnetic films are identified in Landauer [1961] that satisfy the assumptions of time-reversible, conservative operation. The conservative processes can be used to construct elements with switchable stable states to realize bits for computation. The switching is assumed to be deterministic, that is, the state after any given switching operation is unique. The following deterministic switching operations on every bit are reasonably assumed to be physically realizable without dissipating energy: switching from 0 to 1, switching from 1 to 0, staying at 0 while being at 0, and staying at 1 while being at 1. Crucially, conservative operation of a switch is not possible if the current state value for a bit is not known, that is, energy is necessarily dissipated when switching from an unknown bit value to either 0 or 1. Such a switch from an unknown bit value to a specific bit value is not possible to perform conservatively because it violates the time-reversible operation of the conservative process. Otherwise, a time-reversed operation after reaching the specific bit value must result in two different bit values, which contradicts determinism. The switch from an unknown bit value to a specific bit value is called bit erasure. Because it is possible to switch from 0 to 1 or vice versa with zero energy cost, the bit erasure can be defined without loss of generality as resetting an unknown bit value to 0. Landauer placed the lower bound on the dissipated energy per bit erasure as $kT \ln 2$ joules, as mentioned earlier.

From the aforementioned observations, it follows that, in principle, only bit erasures impose a non-zero lower bound on the amount of dissipated energy for any computation. If bit erasures are avoided in a computation, Landauer's Principle says it would be theoretically possible to accomplish such a computation with no dissipation of energy. Now consider any reversible computation whose input bits can be uniquely recovered from its output bits. The set of all reversible computations is exactly the same as the set of all computations that incur no bit erasures. Hence, by Landauer's Principle, reversible computations can be theoretically performed without any dissipated energy. However, historically, conventional computers have been designed for irreversible programs, and hence, for the findings to be relevant, it is necessary to analyze the energy cost of irreversible computations as well. It is in this context that the results of Bennett [1973] and others bear relevance, providing methods by which irreversible computations may be performed over reversible systems, albeit trading off some cost in time and space to avoid energy cost.

In the next sections, the preceding principles are examined in the context of logic circuits and machines, respectively. The former concerns a system-theoretic view of reversible logic circuits. The latter concerns a relaxation of the requirement of deterministic execution assumed in the development of the Landauer's Principle.

4.3   Reversible Circuits as Bit Compressors

In this section, we will examine the implication of Landauer's Principle on a logic circuit view of computations, and how the reversibility and energy concerns are manifested in the operation of a logic circuit.

4.3.1   Irreversible User Circuit within an Expanded Reversible Circuit

With logic circuits, the desired computation is specified as a function $\mathcal{F}$ on input bits $\mathcal{I} = [I_1, \ldots, I_n], n \geq 1$, giving output bits $\mathcal{O} = [O_1, \ldots, O_m], m \geq 1$. Because in general the mapping of the input to output may be irreversible, the function must be embedded in an expanded mapping that can be assured to be reversible by design. For reversibility, the expanded function introduces additional bits $\mathcal{L} = [L_1, \ldots, L_l], l \geq 0$ on the input side, and $\mathcal{D} = [D_1, \ldots, D_d], d \geq 0$, on the output side. Note that $n + l = m + d$, because reversibility requires the number of input bits to be equal to the number of output bits. Due to the fact that the desired function is only concerned with $\mathcal{I}$ and $\mathcal{O}$, the circuit must internally supply all the bits $\mathcal{L}$ and consume all the bits $\mathcal{D}$ for every evaluation of $\mathcal{F}$.

4.3.2   Clean and Dirty Bits

Conventionally, the $\mathcal{L}$ bits are called *clean* bits, and the $\mathcal{D}$ bits are called *dirty* bits. The "cleanliness" of the $\mathcal{L}$ bits means that the value of the bits is fixed, independent of $\mathcal{I}$, and, more importantly, independent of any prior values for previous inputs of $\mathcal{I}$ as well. Similarly, the "dirtiness" of the $\mathcal{D}$ bits, as counterparts, means that the values of $\mathcal{D}$ are dependent on actual values for $\mathcal{I}$ given by the user. Thus, the circuit is tasked with the job of converting input-dependent $\mathcal{D}$ into input-independent $\mathcal{L}$ between two consecutive computations of $\mathcal{F}$. This task of conversion requires erasure (and eventually resetting) of $\mathcal{D}$ into $\mathcal{L}$. Here we can assume that $\mathcal{D}$ only contains dirty bits (i.e., whose values vary with $\mathcal{I}$), and does not contain any clean bits. This is because any clean bits in $\mathcal{D}$ can only originate from $\mathcal{L}$.

   The circuit implementation is not obligated to perform the conversion of $\mathcal{D}$ to $\mathcal{L}$ soon after every evaluation, and may in fact choose to postpone conversion to be performed in batches. However, it cannot postpone every conversion indefinitely because of the physical resource constraints (e.g., bounded physical size) of the circuit. Nevertheless, the compression may be logically considered to be carried out after every reversible function evaluation.

4.3.3 Custom Computation Circuit

The design of a custom circuit C^s using these principles is illustrated in Figure 4.3. The functionality is separated into two distinct components of the user circuit: one called the *Core circuit* C_c^s and the other called the *Compressor circuit* C_r^s. The core circuit captures the functional aspect for the reversibility of $\mathcal{F}$. The compressor circuit logically captures the operating functionality, namely the generation of clean bits and consumption of dirty bits. Note that the external interface to the circuit is only in terms of $\mathcal{I}$ and $\mathcal{O}$, conforming to the user specification. Additional *control* circuitry is needed (e.g., timing circuitry is not shown) to synchronize and/or choose the direction of execution ($\mathcal{F}$ or $\mathcal{F}^{-1}$).

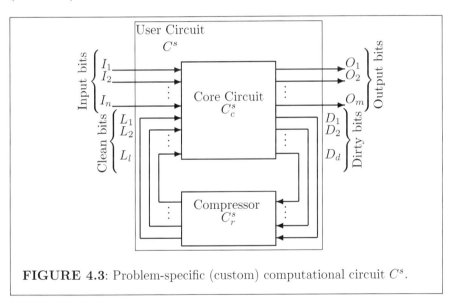

FIGURE 4.3: Problem-specific (custom) computational circuit C^s.

4.3.4 General-Purpose Computation Circuit

To design a general-purpose circuit C^g, the custom circuit template of C^s can be extended to include as input an additional *program* bit vector $\mathcal{P} = [P_1, \ldots, P_p], p \geq 1$. For reversibility, $n + p + l = m + d$. This is shown in Figure 4.4. At a conceptual level, these can be included in the input vector $\mathcal{I}$, but the logical separation helps distinguish the intended purposes of the two sets of vectors.

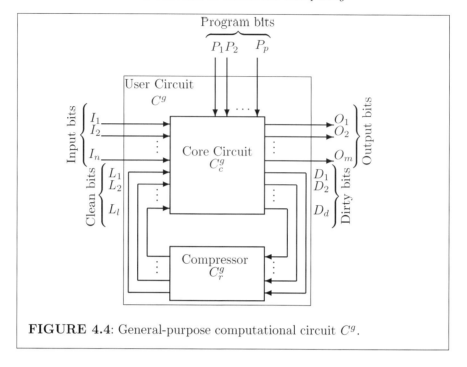

FIGURE 4.4: General-purpose computational circuit C^g.

4.3.5   Energy Cost of the Circuit

Landauer's Principle says that an unavoidable energy loss occurs when d bits are returned to the compressor and converted to l clean bits. This is because the conversion is a bit erasure operation on the dirty bits. Hence, when a circuit is embedded into an expanded reversible circuit that results in d bit erasures to be performed for every circuit evaluation, the bit compressor at temperature T Kelvin necessarily dissipates at least $kT \ln 2^d$ joules per cycle.

4.3.6   Analogy with Refrigeration

An interesting analogy exists between the compressor in a refrigerator and the notion of a *bit compressor* in a reversible logic circuit. Just as the refrigerator compressor cannot be made to convert hot refrigerant to a cooler one with 100% thermodynamic efficiency, so too the bit compressor cannot be made to convert dirty bits to clean bits with 100% thermodynamic efficiency. Also, just as the refrigerator needs a compressor for its operation because it cannot be supplied with an unlimited amount of cool refrigerant, the reversible machine needs a bit compressor because it cannot be supplied with an unlimited amount of clean bits. The core circuit in the logic circuit, however, does not suffer from any fundamental limitations on efficiency and can be made arbitrarily high, in exchange for an increase in the time taken to compute and/or the amount of intermediate state (space) used. This is analogous to the possibility of making the cooling operation itself (namely, achieving and maintaining the temperature differential between the exterior and the interior of the refrigerator) as efficient as one desires so long as there is a continuous supply of cool refrigerant or the cooling process is allowed to proceed sufficiently slowly.

4.3.7 Reversibility in the Eye of the Beholder

Let us denote the computation of the user-specified irreversible circuit by the functional relation $\mathcal{O} = \mathcal{F}(\mathcal{I})$, where $\mathcal{I}$ is the vector of input bits and $\mathcal{O}$ is the vector of output bits, as described before. The reverse evaluation of $\mathcal{F}$ is defined as the recovery of $\mathcal{I}$ from $\mathcal{F}(\mathcal{I})$.

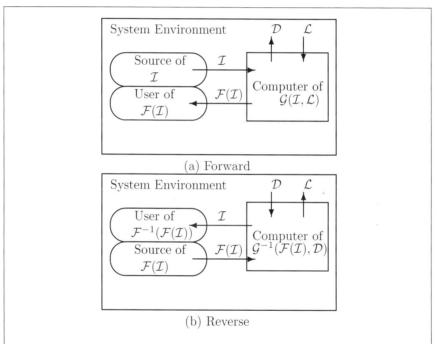

FIGURE 4.5: Relation between the bits of interest to the user and the actual set of bits *reversibly* computed.

4.3.7.1 Reversal via Expanded Function

Because in general $\mathcal{F}$ may not be invertible, $\mathcal{F}$ is embedded inside a new functional relation $(\mathcal{O}, \mathcal{D}) = \mathcal{G}(\mathcal{I}, \mathcal{L})$ defined as shown in Figure 4.5(a). The input $\mathcal{L}$ can only be recovered by inverting $\mathcal{G}$ instead of inverting $\mathcal{F}$ alone. In the simplest case, $\mathcal{F} = \mathcal{G}$ and $\mathcal{L} = \mathcal{D} = \emptyset$, but this is not true in general.

The function $\mathcal{F}^{-1}$ that recovers $\mathcal{I}$ from $\mathcal{F}(\mathcal{I})$ is obtained indirectly by relying on the reversibility of the expanded function $\mathcal{G}$, as shown in Figure 4.5(b). The inverse function $\mathcal{G}^{-1}$ for $\mathcal{G}$ has the property

$$(\mathcal{I}, \mathcal{L}) = \mathcal{G}^{-1}(\mathcal{O}, \mathcal{D}).$$

Among all possible $\mathcal{G}$ for a given $\mathcal{F}$, let $\mathcal{G}^*$ be an optimal function that minimizes the number of dirty bits produced, that is, $d^* = |\mathcal{D}^*|$ produced by $\mathcal{G}^*$ is the smallest among all $\mathcal{G}$. Reversible computing techniques deal with realizing a $\mathcal{G}^*$ corresponding to any given $\mathcal{F}$. Importantly, this problem of minimizing the number of dirty bits only arises from the specific nature of $\mathcal{F}$ as defined *by the user*.

In particular, if indeed the user were to be *interested in* a new function

$$\mathcal{O}' = \mathcal{H}(\mathcal{I}'),$$

where $\mathcal{I}' = \mathcal{I} \cup \mathcal{L}$, $\mathcal{O}' = \mathcal{O} \cup \mathcal{D}$, and $\mathcal{O} = \mathcal{F}(\mathcal{I})$, then the new function of interest, $\mathcal{H}$, is perfectly reversible without the need to expand into a different reversible function $\mathcal{G}$. In the reversible computation of $\mathcal{H}$, no clean bits are needed from the system, and no dirty bits are pushed into the environment.

4.3.7.2   Relating Energy and User Interest

Given the function $\mathcal{F}$ and the corresponding $\mathcal{G}^*$, the minimum energy dissipated in computing $\mathcal{F}(\mathcal{I})$ with perfect efficiency is given by $E_{min} = kT \ln 2^{d^*}$ due to the fact that d^* bits are erased at temperature T. A key observation is that the dissipated energy is due to information that is being discarded, apparently from *disinterest* on the user's part. The disinterest is manifested implicitly in the definition of the irreversible function $\mathcal{F}(\mathcal{I})$ specified by the user. On the other hand, if the user had been willing to expand his interest from $\mathcal{F}$ to its corresponding function $\mathcal{H}$, the computation can be theoretically made to operate with zero dissipated energy because $\mathcal{H}$ generates no dirty bits. In other words, any energy used in the (forward) computation of $\mathcal{H}$ is possible to be recovered without loss in the computation of $\mathcal{H}^{-1}$. This variation of the minimum dissipated energy with user's specific interest makes the effects of reversibility dependent on the eye of the beholder, or on the user's choice in the demarcation of the system boundaries.

4.4   Deterministic versus Non-Deterministic Reversal

Here we revisit Landauer's Principle, focusing on the assumption of determinism that underlies its proposition, and examine an alternative view in which non-determinism due to irreversible computation is accepted. An important outcome of this relaxation is that the energy cost of bit erasures may be overcome not only by methods that avoid bit erasures but also by a different method that exploits non-determinism.

4.4.1   Bit Erasure Cost versus Bit Reset Cost

Recall that the energy cost of $kT \ln 2$ per bit erasure as stated by Landauer's Principle is based on the requirement of deterministic computation. Maroney revisited the analysis and offered certain clarifications to the principle's scope, interpretation, and generalization [Maroney, 2004, 2005, 2009]. Maroney pointed to the importance of the subtle difference between a bit *erasure* and bit *reset* operation. When a bit is being "forgotten" as part of a computation, it is the *reset* operation on that bit, and not the *erase* operation on the bit's current value, that necessarily *dissipates* energy. Bit erasure is the act of losing track of the bit's current value. Bit reset takes the unknown bit value and sets it to a known value (say, 0). Hence, in a more precise understanding of the energy cost of logical computation, a distinction must be made between the energy costs of the two operations. Maroney observed that the bit erasure does not require dissipation of energy, but it is only a subsequent *deterministic* bit reset that incurs that cost. Moreover, after a bit has been erased, it may be *non-deterministically* reset (to a value dynamically determined by the current state of the system) such that the net energy cost for the bit erasure and non-deterministic reset is zero. In light of Maroney's correction and generalization, the energy cost in Landauer's Principle must be qualified as *deterministic bit reset cost*, as opposed to a *bit erasure* cost. This distinction leads to two different ways of executing irreversible computations with zero energy cost. Both deal with the ability to produce the output of arbitrary programs with no dissipated energy. However, their difference only lies in the restoration of the input, based on the distinction between deterministic or non-deterministic execution, as elaborated next.

4.4.2   Zero Energy Cost Schemes

In general, a computation can be deterministic or non-deterministic in the forward execution. The same computation may also be deterministic or non-deterministic in the backward direction. Thus, there are four types of computation, as shown in Table 4.1. A program that has non-determinism in backward execution is conventionally called an irreversible program. Nevertheless, since the notions of forward and reverse are interchangeable, the notions of (ir)reversibility and (non-)determinism may be equally applied to either forward or reverse mode of execution.

The notions of irreversibility and non-determinism of any state sequence are illustrated in Figure 4.6. Each black disk represents a state. An arrow from state s to state d represents that the execution of the program takes the

TABLE 4.1: Types of Computation in Terms of Reversibility and Determinism

Code	Forward Execution	Backward Execution
T1	Non-deterministic	Deterministic
T2	Deterministic	Deterministic
T3	Deterministic	Non-deterministic
T4	Non-deterministic	Non-deterministic

state from s to d. In the figure, irreversibility corresponds to $n > 1$, and non-determinism corresponds to $m > 1$. Again, note that the notions of irreversible and non-deterministic evolutions in the figure can be applied to both forward and reverse modes.

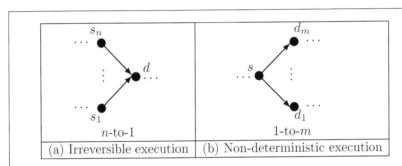

FIGURE 4.6: Irreversibility and non-determinism in state updates.

With the preceding background, let us examine the two different ways of executing any computation $\mathcal{O} = \mathcal{F}(\mathcal{I})$ with zero energy cost.

1. **Bennett's scheme**: The first method is part of the previously mentioned result that showed that any computation can be performed with zero bit erasures. Using this method, the system starts with the original input $\mathcal{I} = \mathcal{I}_{in}$ and terminates with the pair $(\mathcal{I}_{out}, \mathcal{O})$, where $\mathcal{O}$ is the output of the computation and $\mathcal{I}_{out} = \mathcal{I}$ is the same as the original input. Determinism is assumed as a requirement of execution; as a result, $\mathcal{I}_{out}$ must be equal to $\mathcal{I}_{in}$ even if $\mathcal{F}$ is not one-to-one. To ensure deterministic execution, this method generates an intermediate log of bits that are reversibly erased before the computation terminates. Hence, extra bit space (and some extra operations) are needed in this method.

2. **Maroney's scheme**: This method exploits the possibility of non-deterministic execution to obviate extra space and time needed in the Bennett's scheme. The computation is performed to produce the output $\mathcal{O}$ on termination. However, unlike Bennett's scheme, the copy of the

input $\mathcal{I}_{out}$ remaining at the end of computation is not necessarily equal to the original input, but can be potentially any other input that gives the same output, that is, $\mathcal{F}(\mathcal{I}_{out}) = \mathcal{O}$.

4.4.2.1 Bennett's Scheme

To show that computation can be (asymptotically) performed with zero energy cost, Charles Bennett developed an ingenious scheme that takes any conventional irreversible program and theoretically executes it with no bit erasures. The scheme is a "compute–copy–uncompute paradigm," illustrated in Figure 4.7 (see also Section 7.2), which works as follows. The program is first executed on the input to generate the output. This can be performed without energy dissipation by keeping a record of all activity so that no bit values are lost (thereby avoiding susbsequent bit erasures). A copy of the output is made to save the output. The entire forward computation is then undone by executing the program backward, thereby restoring all bits to their original values. Because the net dissipated energy is zero, $E_{in} = E_{out}$. For this scheme to work, the original program must be reversible to be able to execute it in the reverse mode, but we must assume that the program can be irreversible. Hence, the original irreversible program must first be converted to a semantically equivalent reversible program. This can be achieved by logging information during execution. When this converted program is executed in the forward mode, a set of "clean" bits is needed to record the runtime log in a reversible manner. These clean bits are converted to dirty bits at the end of the forward execution. The reverse mode of the (modified) program restores them to their clean state.

A crucial observation about this scheme is that both the forward and reverse executions are required to be *deterministic*. This important assumption or reliance on deterministic execution ensures that the entire execution not only generates the output but also restores the input to exactly the same value it had before the program started. In other words, it retains a one-to-one mapping between the input and the output.

4.4.2.2 Maroney's Scheme

It is also possible to compute with zero energy cost by relaxing the deterministic execution requirement in the reverse mode [Maroney, 2004]. The only change to the semantics of the reversible execution is that the input is now no longer guaranteed to be the same as the original input. However, it does guarantee that the restored input is in fact another valid input that would give the same output as the original input. This can be easily expected because irreversibility of the original program essentially means that there is more than one valid input to the program that generates the same output. The new scheme is shown in Figure 4.8. In this, the original program is executed in the forward mode unmodified. The reverse of the program is allowed to contain non-determinism; hence no special instrumentation of the program is needed.

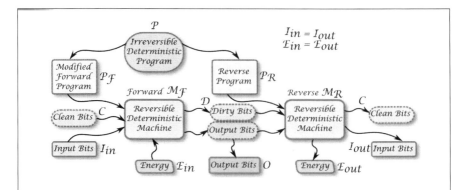

FIGURE 4.7: Bennett's scheme to incur zero energy cost in executing an irreversible deterministic program on a reversible *deterministic* machine.

This approach also obviates the notions of clean and dirty bits. Just as in Bennett's scheme, net energy loss, $E_{in} - E_{out}$, is zero.

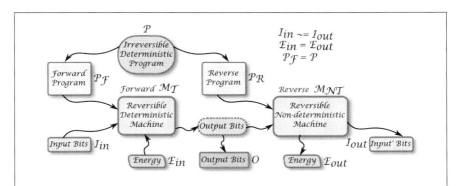

FIGURE 4.8: Maroney's scheme to incur zero energy cost in executing an irreversible deterministic program on a reversible *non-deterministic* machine.

Chapter 5

Reversibility-Related Paradoxes

5.1 Entropy

Any discussion about reversibility inevitably touches on another important concept, namely that of *entropy*. The entropy of a system is often loosely defined as a measure of "disorder," although such a definition is imprecise because disorder is a subjective concept. In reality, in a closed dynamical system, any increasing function (or, more generally, any non-decreasing function that

49

"eventually" increases) of the system's variables that can be identified in that system can be defined as an entropy function of that system. Functions that never increase (i.e., which stay constant, because they also do not decrease) are not interesting entropy functions, especially if at least one other entropy function can be found that experiences some increase. To identify and/or define an entropy function of a system, the system is viewed as a composition of two or more subsystems. The values of variables representing the enclosed subsystems are called microstates, and the aggregations of the enclosing system are called macrostates. An entropy function of the system is then determined as a function of possible mappings from macrostates to microstates as the system evolves. When the system moves from an "old" macrostate to a "new" macrostate, a potential ambiguity arises about the specific set of microstates that underlie the old macrostate that gave rise to the new macrostate. In general, if there is nothing in the system that can be used to resolve this ambiguity, then the system accumulates this ambiguity about the previously taken path during its evolution. Any function that captures this increase of ambiguity about the system's evolution path qualifies as *an* entropy function of the system [Gottwald and Oliver, 2009].

5.2 Reversibility and Entropy

Historically, the concepts of reversibility and entropy were considered to be tightly interlinked. An interdependence was assumed to exist between reversibility of system evolution and the increase in entropy of the system, resulting in various implications and expectations. For example, if the steps in a system's evolution are individually reversible, it could be (incorrectly) expected that there would be no necessary increase in entropy. Conversely, an increase in entropy could be (incorrectly) expected to imply that there is some source of irreversibility in the system evolution. In addition to such considerations with respect to the evolution of any dynamical system, new considerations apply to computation or simulation of such systems. For example, when simulating a phenomenon in which an entropy measure exists, it might be (incorrectly) expected that the increase in entropy would necessarily manifest itself as an increase in memory that would be necessarily accumulated in a reversible execution of that simulation.

All such incorrect expectations result from an incorrect assumption of an allegedly necessary relation between the concepts of reversibility and the concept of entropy. The independence of reversibility and entropy would be clear if there are examples in which the system remains reversible at every step of evolution and yet exhibit an increasing entropy measure (or, equivalently, an "arrow of time" in the system evolution). Two commonly used examples available in the literature are described in the next sections to illustrate this independence between reversible execution and entropy increase (and, thereby, independence of entropy increase and the memory needed to perform reversible simulations). The first example is the Ehrenfest's Urn model, and the second example is the Kac-Ring model. Both clearly exhibit perfect reversibility that results in zero memory increase despite indefinite lengths of execution, yet in both examples there exist increasing entropy functions. In particular, the Ehrenfest's Urn model shows a clear one-way evolution from any initial state to an equilibrium state, that effectively gives an arrow of time for the system; yet the system can be reversibly executed without any growing memory trace. While both examples illustrate how irreversibility arises at a macro level even though the systems remain reversible at the micro level, here we use them to illustrate the reversibility with respect to computation (or simulation) on the computer, and show how a reversible execution could be memory-less despite the presence of an increasing entropy for the simulated system.

5.3   Ehrenfest's Urn Model

The *Ehrenfest's Urn* model is one of the simplest systems designed to understand the relation between reversibility and increase of entropy. This example has been used in the literature to clarify the apparent paradox of one-way evolution always arising out of an otherwise reversible phenomenon. It can serve to discuss concepts such as entropy increase and an arrow of time. Here, we will also use this example to show how the classical view of reversibility (e.g., in physics) is different from the view needed for computer-based reversible simulation of a physical phenomenon.

5.3.1   Model Configuration and Operation

The model consists of two urns and N numbered balls, as illustrated with $N = 10$ in Figure 5.1. Initially, the N balls are divided across the two urns, with N_1 balls in the first urn and $N - N_1$ in the other. In the initial conditions of interest, $N_1 \ll N$. Due to the symmetrical and complementary nature of the two urns, the model can focus on tracking the occupancy of only one of the urns, say, the first urn. In each step, an N-way random number, $R_N \in [1, N]$, is thrown, and the ball numbered R_N is moved from its current urn to the other urn. As the number of steps increases, the number of balls in the first urn changes. The function of interest is the probability, $P_j(t), 1 \leq j \leq N$, that the ball numbered j would be found in the first urn after t steps. Clearly, this probability distribution changes as t increases. Also of interest is the probability distribution, $Q_j = \lim_{t \to \infty} P_j(t)$, that makes it independent of time and gives the steady-state distribution.

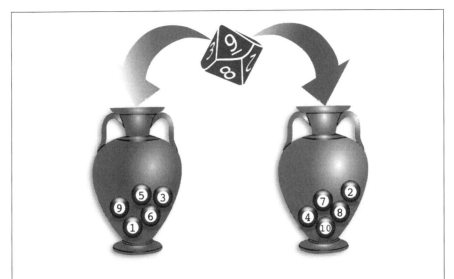

FIGURE 5.1: The Ehrenfest's Urn model illustrated with $N = 10$ balls.

5.3.2 Analysis

Let the pair $(j, N - j)$ represent the state of the system at some time step, to denote that the first urn contains j balls and the second contains the rest. Then, the behavior of the system can be modeled as a Markov chain of transitions among the states, as shown in Figure 5.2. With an N-faceted die, the probability of transition *per ball* is $\lambda = \frac{1}{N}$. The state transitions from $(j, N - j)$ to $(j + 1, N - j - 1)$ with probability $(N - j)\lambda$ because there are $N - j$ balls in the second urn that are candidates for a ball to move from the second urn to the first urn. Similarly, the state transitions from $(j, N - j)$ to $(j - 1, N - j + 1)$ with probability $j\lambda$ because there are j possible balls out of N currently in the first jar from which one ball can be moved out.

Starting with the initial conditions (e.g., starting with the state $(N, 0)$), the system goes through transitions until it smears itself probabilistically, so to say, throughout the entire Markov chain such that every state has been visited very many times and historical effects become less and less relevant in determining the probability associated with any state. In such a state, called the *equilibrium*, a time-independent probability distribution of the states can be determined by calculating the number of possible ways in which any given state can be composed out of all the possibilities. This can be computed as follows.

Consider an N-bit integer formed by setting its bth bit to 0 if ball b is in the left urn, 1 otherwise. Because any ball can be in either the left urn or the right urn, the integer can take on all the values from 0 to $2^N - 1$. Thus, there are $C = 2^N$ possible configurations of the system. There are $c = \binom{N}{j}$

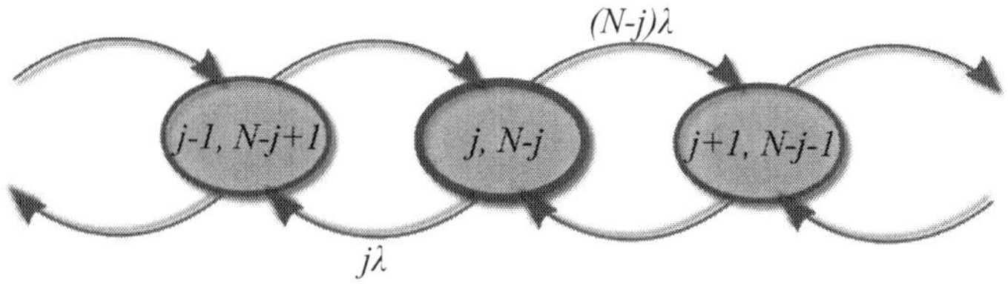

$$
\begin{aligned}
j &= 0, \dots, N \\
\lambda &= \tfrac{1}{N} \\
P_j(t) &= \text{Probability system is in state } (j, N-j) \text{ at time } t \\
Q_j &= \text{Probability system is in state } (j, N-j) \text{ as } t \to \infty \\
&= P_j(\infty) = \tfrac{1}{2^N}\binom{N}{j}
\end{aligned}
$$

FIGURE 5.2: Markov chain and probability distributions in the Ehrenfest's Urn model.

ways in which j balls can be chosen from N balls, all of which are the same configuration when they are in the same urn, which occurs in one out of C configurations. Hence, the probability of finding j balls in the left urn in the equilibrium state is $Q_j = \frac{c}{C} = \frac{1}{2^N}\binom{N}{j}$.

5.3.3 Forward and Reverse Algorithms

The normal (forward) operation of the urn model and its reverse operation algorithm are shown in Algorithm 5.1. The evolution processes are described in the form of computer programs that can be executed to simulate the system. The system is configured using the **Initialize** routine. Invocation of the **Forward** routine followed by the **Reverse** program will restore the system exactly to its initial state. This condition is verified after the reversal via assertions in the **Verify** routine. The availability of a reversible random number stream is assumed (see Chapter 12), from which samples are drawn to initialize the system with N balls randomly assigned to the two urns. The same stream is used after initialization to generate random identifiers of the balls to be moved between the urns.

5.3.4 System Reversibility versus Computational Reversibility

Consider the function

$$
H(t) = -\sum_{j=1}^{N} P_j(t) \log \frac{P_j(t)}{Q_j}.
$$

Algorithm 5.1 Forward and reverse algorithms for the Ehrenfest's Urn model.

N=number of balls
T=number of operations (random swaps) performed
$\mathcal{R}(i..j)$=random integer between i and j inclusive
$\mathcal{R}^{-1}(i..j)$=reverses the random stream, recovers recent $\mathcal{R}(i..j)$
$U[]$=array of N bits
$U[b]$=0 if ball b is in left urn, 1 if in right urn

Initialize
for $b = 1$ **to** N **do**
$U[b] \leftarrow \mathcal{R}(0..1)$
end for

Forward	Reverse
for $t = 1$ **to** T **do**	**for** $t = T$ **down to** 1 **do**
$b \leftarrow \mathcal{R}(1..N)$	$b \leftarrow \mathcal{R}^{-1}(1..N)$
$U[b] \leftarrow 1 - U[b]$	$U[b] \leftarrow 1 - U[b]$
end for	**end for**

Verify
for $b = N$ **down to** 1 **do**
assert $U[b] = \mathcal{R}^{-1}(0..1)$
end for

This function is an entropy measure for this system because it can be shown[Kelly, 1979] that $H(t)$ is strictly increasing as $t \to \infty$.

At every t, for reversibility, the system requires us to record the deviation of the actual distribution $P_j(t)$ from the equilibrium distribution Q_j. This record will need, on average, $h_j(t) = \log P_j(t) - \log Q_j = \log \frac{P_j(t)}{Q_j}$ bits for each possible j with probability $P_j(t)$. Hence, on average, $H(t) = \sum_{j=0}^{N} P_j(t) \cdot h_j(t)$ bits are needed to remember the actual trajectory taken by the system, starting from any given initial condition to the current time t. Thus, the system's entropy measures the loss of information over time with respect to system reversibility. *By contrast, there is no loss of information with respect to computational reversibility: the steps can be retraced deterministically and without additional information.* With pseudorandom number streams, reversal of the random number stream can be realized with zero memory cost. However, with externally sourced streams, there is a bit cost incurred in remembering the forward stream in order to go back in reverse. The bit cost then is equal to $t \cdot w$, where w is the bit precision of the random number.

It is clear that increasing entropy does not imply irreversibility, and that the reversibility aspects of the system are different from those for a reversible *simulation* of the system. Thus, although there is an "arrow of time" in the evolution of the *physical* system, it does not interfere with the *computational* reversibility of a simulation of the system.

5.4 Kac-Ring Model

Another model that illustrates the difference between microscopic reversibility and macroscopic entropy is the Kac-Ring model [Kac, 1956]. In the following description, we borrow the common notation from the literature [Gottwald and Oliver, 2009].

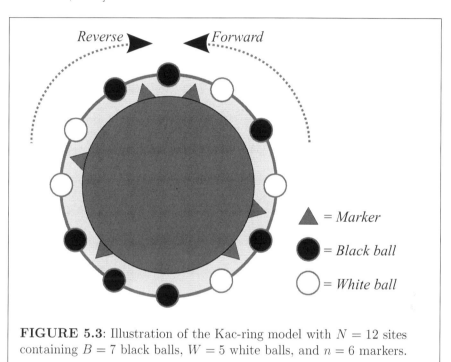

FIGURE 5.3: Illustration of the Kac-ring model with $N = 12$ sites containing $B = 7$ black balls, $W = 5$ white balls, and $n = 6$ markers.

5.4.1 Model Configuration and Operation

A set of N balls is arranged in a circular fashion. Each ball is either white or black in color. A wheel with n spokes turns inside the circular ring of balls. Each spoke is called a marker, and the n markers are randomly placed in the N slots along the circumference of the wheel. An example is shown in Figure 5.3 for a small number of balls and markers. Whenever a marker goes past a ball, the ball's color is toggled (white becomes black, black becomes white). If each turn of the wheel constitutes one time unit of evolution, the object of interest is the state of the system after t turns, or after time t. The following observations can be made about the system:

1. If n is even, the system reverts back to the initial configuration after $t = N$. If n is odd, the initial configuration is reached after $t = 2N$.

2. Reversibility is easily achieved: forward evolution is obtained by rotating the marker wheel counterclockwise while reverse evolution is obtained simply by flipping to clockwise rotation

3. In general, as $N \to \infty$, it takes $t \to \infty$ for the system to revisit its initial state.

5.4.2   Analysis

Let the number of white balls at time t be W_t. The number of black balls, B_t, is equal to $N - W_t$. Among the W_t white balls, let w_t balls be toggled by the turn of the markers at time t; similarly, let b_t of the B_t black balls be toggled. For example, in the forward (counterclockwise) mode shown in Figure 5.3, there are $b = 4$ black balls about to be toggled, and $w = 2$ white balls about to be toggled. It follows that $W_{t+1} = W_t + b_t - w_t$, and $B_{t+1} = B_t + w_t - b_t$. Let $\Delta_t = B_t - W_t$. Then $\Delta_{t+1} = B_{t+1} - W_{t+1} = (B_t - W_t) - 2(b_t - w_t) = \Delta_t - 2(b_t - w_t)$. This implies that the difference between the number of black balls and the number of white balls depends only on w_t and b_t, which is the microscopic information about precisely which balls are positioned at the markers about to be toggled next.

Now consider a random configuration of a ring with N balls, in which N is very large. Assuming uniformly random placement of markers and ball colors, the ratio of markers to balls is the same as the ratio of the white balls in front of markers to the total number of white balls (and analogous ratios for black balls as well). Thus, if the ratio is μ, then

$$\mu = \frac{n}{N} = \frac{w_t}{W_t} = \frac{b_t}{B_t}. \tag{5.1}$$

On average, if the same argument about uniformity is accepted at every t, the ratios can be used in the recurrence relations, giving the expected value $\widetilde{\Delta}_t$ of Δ_t as $\widetilde{\Delta}_{t+1} = (1 - 2\mu)\widetilde{\Delta}_t = (1 - 2\mu)^{t+1}\Delta_0$. Thus,

$$\widetilde{\Delta}_t = (1 - 2\mu)^t \Delta_0.$$

Interestingly, Equation (5.1) is analogous to the *Stosszahlansatz* condition that Boltzmann used in his H-theorem for molecular disorder (see, for example, [Lebowitz, 1994]). In modeling the aggregate behavior of a large number molecules in an ideal gas, in addition that the pre-collision velocities of molecules are uncorrelated, Boltzmann introduced an assumption that post-collision velocities are also uncorrelated. This additional assumption about post-collision velocities made the calculations of the system dynamics possible. By a similar analogy, Equation (5.1) assumes that the post-rotation ratios of white to black balls remains uncorrelated, although they are correlated due to the specific configuration of markers and the initial conditions of the ring.

5.4.3   Forward and Reverse Algorithms

The normal (forward) operation of the Kac-Ring model is shown in Algorithm 5.2. The operation is described in the form of computer programs that simulate the system, given any specific input value for N. The system is configured and initialized with the **Initialize** routine. The **Forward** routine evolves the system forward by T steps, after which the **Reverse** routine evolves the system backward by T steps to bring the system back to its initial state. Perfect restoration via reversal is checked in the **Verify** routine that should be invoked after reversal.

5.4.4   An Entropy Function

Consider the function

$$S(\Delta_t) = \frac{1}{N} \log C_t, \text{ where } C_t = \binom{N}{B_t}.$$

This function defines, on average per ball, the logarithm of the number of distinct microscopic configurations of the Kac-ring of N balls that give the same macroscopic configuration as specified by a given value for Δ_t. The number of micro states is nothing but all the possible arrangements of B_t black balls and $W_t = N - B_t$ white balls in the ring. This is equal to the number of ways in which N balls can be partitioned into two sets: one set with B_t balls and the other with $W_t = N - B_t$ balls. This is equal to $C_t = \binom{N}{B_t}$. To identify precisely which one of these C_t configurations was in fact the predecessor of the current macro state, the number of bits needed is equal to $\log C_t = N \cdot S(\Delta_t)$.

Let $\delta_t = \frac{\Delta_t}{N}$. Then, using the Stirling's approximation $\log x! \cong x \log x$,

$$S(\Delta_t) = 1 - \frac{1}{2}\left[(1 + \delta_t)\log_2(1 + \delta_t) + (1 - \delta_t)\log_2(1 - \delta_t)\right].$$

This implies that $S(\Delta_t)$ increases from 0 to 1 as δ_t decreases (from an initial value $x \leq 1$) down to 0 because Δ_t decreases (from an initial value $X \leq N$) down to 0. Because Δ_t decreases with t, $S(\Delta_t)$ increases with t. Thus, $S(\Delta_t)$ defines an entropy function for the system that increases as the system evolves with increasing t.

5.4.5   System Reversibility versus Computational Reversibility

The existence of the preceding increasing entropy function follows from the Stosszahlansatz condition, which itself is considered a reasonable expectation when N is very large. Historically, philosophical debates followed this line of reasoning, but for present purposes, it suffices to note that the system exhibits

Algorithm 5.2 Forward and reverse algorithms for the Kac-ring model

N=number of balls
B=number of black balls
W=number of white balls
n=number of markers
$M[]$=array of N bits
$M[i]$=1 if a marker is in place i, 0 otherwise
$A[]$=array of N bits
$A[i]$=1 if ball i is white, 0 if black
$\mathcal{R}_b$=next bit from random bit stream
$\mathcal{R}_b^{-1}$=reverses the bit stream, recovers recent $\mathcal{R}_b$
T=number of time steps
x **XOR** y=exclusive-or of bits x and y

Initialize
$B \leftarrow 0$
$n \leftarrow 0$
for $i = 1$ to N do
$\quad M[i] \leftarrow \mathcal{R}_b$
$\quad$ if $M[i] = 1$ then $n \leftarrow n + 1$
end for
for $i = 1$ to N do
$\quad A[i] \leftarrow \mathcal{R}_b$
$\quad$ if $A[i] = 0$ then $B \leftarrow B + 1$
end for
$W \leftarrow N - B$

Forward	Reverse
for $t = 1$ to T do	for $t = T$ down to 1 do
$\quad$ for $i = 1$ to N do	$\quad$ for $i = N$ down to 1 do
$\quad\quad j \leftarrow 1 + ((i + t - 1) \mod N)$	$\quad\quad j \leftarrow 1 + ((i + t - 1) \mod N)$
$\quad\quad A[i] \leftarrow A[i]$ **XOR** $M[j]$	$\quad\quad A[i] \leftarrow A[i]$ **XOR** $M[j]$
$\quad$ end for	$\quad$ end for
end for	end for

Verify
for $i = N$ down to 1 do
$\quad$ assert$(A[i] = \mathcal{R}_b^{-1})$
end for

a one-way evolution of the dynamics despite trivially reversible mechanics of the underlying process.

Recapitulating all the important aspects of the model, we see that the system is fully deterministic and reversible at every step of its evolution, and yet it exhibits an analogy of the arrow of time by which an apparent one-way evolution of a macroscopic quantity is evident as the system evolves. All this is seen notwithstanding the fact that it is a closed system with no infusion/emission of information into/from the system. This relation between reversibility and entropy increase is well known in physics. However, the important aspect for reversible computing is that a *computer-based simulation* of a physical phenomenon in which entropy increases does not necessarily make the computation irreversible, nor does it necessarily result in a memory trace for reversible execution.

5.5 Relation to Maxwell's Demon

A hypothetical construction called the Maxwell's Demon is a well-known proposition that attempts to break the view that entropy must always (eventually) increase in a dynamical system. Besides its implications on our understanding of physics in general, it is an excellent illustration of the strange relationship between irreversible (memoryful) computation and entropy in some dynamical systems.

5.5.1 Development

The Maxwell's Demon is a theoretical system that was essentially designed as a counter-argument to the Second Law of Thermodynamics. The Second Law, widely considered as an important observation about the entire universe, effectively states that the thermodynamical entropy of any closed dynamical system increases over time. The Maxwell's Demon *apparently* violates the law; the challenge is to pinpoint where and how the essential discrepancies arise, and see if the paradoxical discrepancies can be effectively resolved. Maxwell's Demon comes in different flavors and variants. A system, for example, was constructed such that, apparently, heat can be made to indefinitely flow from cooler to warmer portions of a system without expending mechanical work or energy, which is in direct violation of the Second Law. Interestingly, it transpires that this paradoxical Demon bears a strong relation to both physical reversibility as well as computational reversibility. Although the physical aspects of the Demon have been widely debated for more than a century, it was only in the mid to late 1900s that a computational side of the Demon was uncovered, bringing with it the relation to reversible computing.

In 1961, Rolf Landauer argued [Landauer, 1961] that the Demon's operation is not without irreversible energy cost, contrary to the original implicit view that the Demon's logical actions do not incur any thermodynamical increase in entropy of the system. He argued that the Demon merely moves its operating energy cost from the visible and obvious physical system to the invisible and intangible cost of realizing reversible logical decisions. Later, Jeffrey Bub [Bub, 2001] and Charles Bennett [Bennett, 2003] extended the reasoning and provided clarification that the irrecoverable energy cost of the logical operations by the Demon arises from the *necessarily blind* erasures of bits of information by the Demon (i.e., fundamentally unavoidable discarding of information stored as logical states). Under the reasonable assumption that the Demon possesses limited memory (or, alternatively, can rely only on finite physical resources for memory), the Demon is eventually forced to forget some of its actions over a sufficiently long course of its operation, which necessarily results in loss of information from the system. The loss of information manifests itself as irreversible bit operations (or, equivalently, as bit erasures), and hence necessarily generates heat.

5.5.2   Setup and Operation

To understand the arguments at a high level, consider the operation of the simplified skeleton of the Maxwell's Demon illustrated in Figure 5.4. The system consists of a set of n particles $S = \{P_i | 1 \leq i \leq n\}$ contained in a box that is divided by a wall into two chambers, called the left chamber, L, and the right chamber, R. A door on the dividing wall is assumed to be operable by the Demon with zero energy dissipation (e.g., with frictionless operation). The act of opening or closing is assumed to be instantaneous. Based on widely accepted feasibility regarding physical realizability, it is also assumed that the cost of measurement of the system state by the Demon may be rendered negligible via appropriate constructions of devices.

5.5.3   Operation as a Computer Program

The essential functionality of the Demon can be distilled into repeated execution of a simple act: whenever a particle P_i is about to hit the door from either side, the Demon evaluates a certain condition $C(S, P_i)$ about the system, and, if the condition is satisfied, the door is opened for an infinitesimally small period (to let the particle pass through to the other chamber) and thereafter immediately closed.

The condition $C(S, P_i)$ varies with each variant of this basic Demon template.

■ In a variant called the *Pressure* Maxwell's Demon, the condition lets through only the particles arriving from the left chamber and keeps the door closed for all particles hitting from the right chamber.

■ In another variant called the *Thermal* Maxwell's Demon, the condition

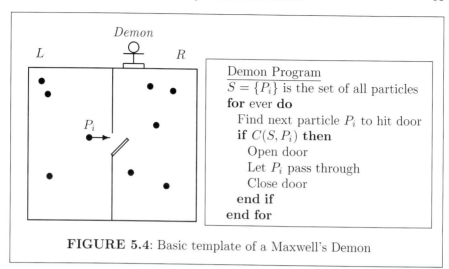

FIGURE 5.4: Basic template of a Maxwell's Demon

evaluates how the approaching particle's velocity compares with the average velocity of all particles. It permits particles from the left chamber moving with a velocity larger than the average velocity to go through; similarly, particles from the right chamber with smaller than the average velocity are permitted as well.

■ In variants such as the Szilard Engine [Szilárd, 1929], the system may contain as few as a single particle, and even the need for accurate measurement of particle state is avoided—the Demon need only know information such as from which chamber the next particle is about to hit the door.

Despite such simplifications, the crux of the problem remains: the Demon must accumulate an infinite amount of information over time, but it cannot retain all of it indefinitely without irreversibly resetting memory states that are realized physically. All the Demon variants are essentially capable of converting the knowledge about the source chamber of the approaching particle into work with zero increase in thermodynamical entropy of the system. Thus, by endowing the Demon with such *intelligence*, perpetual motion appears to become feasible. It is the limitation of this intelligence in terms of memory that is resolved by examining the reversibility of the intelligent computation by any Demon.

5.5.4 Paradox Resolution

When the operation of the Demon is viewed as an automaton, it can be simplified as a computer program shown in Figure 5.4. In the paradox, the

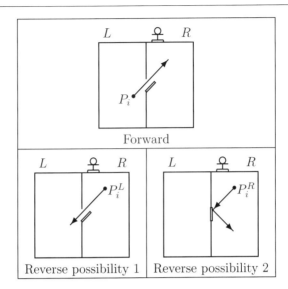

FIGURE 5.5: Illustration of ambiguity for Maxwell's Demon regarding the source chamber of a particle.

Demon can extract work without increasing entropy if and only if the program can be executed reversibly.

Note that, for any iteration, the Demon cannot predict whether the body of the **if** statement would be executed or not executed. Alternatively viewed, the Demon will not be able to run in reverse mode correctly without remembering the past actions because, in general, it cannot disambiguate between two states that result in the same next state. This is illustrated in Figure 5.5. When attempting to reverse the trajectory of a particle P_i that is currently in the right chamber, the Demon is presented with two possibilities for the particle's previous state: either the particle originated from the left chamber and admitted by an open door, or the particle was already in the right chamber and was simply reflected within the right chamber off the closed door. Nothing within the system can offer the information to disambiguate, and the current state is equally probable to occur from both the previous states. This implies that, for every iteration, to be able to reverse the **if** statement, the Demon has to remember as one bit of information (say, 1 to represent that the **if** body is executed, 0 otherwise). Thus, for the program to be reversible, the Demon needs to retain a trace of all the bits recording the sequence of decisions it made. However, with limited memory, the Demon cannot store infinite numbers of bits, and will have to begin overwriting them at some point. Such overwriting of bits irreversibly dissipates energy, that is, irretrievably loses the ability to convert some portion of the energy into useful work. The loss of such an ability to convert energy to work is nothing but an increase in entropy over time. The paradox is thus resolved by observing that information loss appears in the form of irreversible bit erasure.

5.6   Relation to Other Paradoxes

The concept of reversibility arises in a few additional paradoxes, which are briefly described next.

5.6.1   Loschmidt's Paradox

The Loschmidt's Paradox [Boltzmann, 1877, Steckline, 1983] directly brings reversibility arguments to bear on the relationship between microscopic and macroscopic aspects of physical systems. It highlights the apparent disconnect between *reversible* evolutions of *micro*states and *irreversible* evolutions of *macro*states assembled from the microstates.

The basic thesis in this paradox is that it is not possible to derive an irreversible phenomenon at an aggregated (macro) level via a composition of reversible phenomena at the finer (micro) level. If indeed an irreversible macro evolution emerges out of a composition of reversible evolution of microstates of a system, then either there is a flaw in the understanding of the system or the composition itself may have introduced an element of irreversibility that is not present in the microstates. In the most common example, this paradox is illustrated in the context of the kinetic theory of gases: while the motion of particles is reversible, their aggregate properties such as temperature exhibit irreversibility. The Stosszahlansatz condition illustrated in the Kac-ring model (Section 5.4) is an example of the place where composition introduces an element of irreversibility into a phenomenon derived from an aggregation of reversible processes. In the context of reversible computing, the Loschmidt's paradox would imply that a reversible simulation of any phenomenon would accumulate memory in two conditions: either the micro processes underlying the phenomenon are themselves (individually) irreversible, or the specific composition procedure of reversible micro processes itself introduces some type of irreversibility.

5.6.2   Zermelo's Paradox

Zermelo's Paradox [Zermelo, 1896, Steckline, 1983] highlights the apparent contradiction between two different inferences: (1) multiple reversible microscopic processes combining into irreversible macroscopic dynamics that increases entropy, (2) the impossibility of a reversible dynamics to increase entropy due to the fact that the initial state of every reversible process is eventually revisited by the system. The former effect is the type of observation obtained from the one-way evolution of systems such as the Ehrenfest's Urn and Kac-Ring models described in the preceding sections. The latter can be argued as follows. Because every microscopic process is assumed to be reversible, every microscopic state must have a unique predecessor and a unique successor. The uniqueness ensures that no closed loop starting from the initial state can avoid the initial state (because if it did, then there would at least one intermediate state in the loop that has two different predecessors, which is a contradiction of reversibility). Because the initial state will eventually have to be revisited by the system, it implies that entropy cannot always increase, but must begin to decrease at least once in its evolution from the initial state back to the revisit of the initial state. Thus, there is an apparent contradiction.

The apparent contradiction can be resolved by (1) qualifying that the system goes through two phases, first with increasing entropy followed by decreasing entropy, (2) the evolution time of either phase is extremely long compared to any time window of observation; and (3) the system is observable only in the first phase, because of the impossibility of reaching the end of the first phase due to its extremely long duration. Of course, this leaves open questions such as why it is (or, is not) possible to start the evolution at a point in the decreasing entropy phase. While much literature exists on various ways to resolve the paradox (with deep implications to our understanding of physics and universe in general), the relevant observation for reversible computing is that a memory-less, reversible simulation of a composition of reversible microscopic models is indeed possible despite the presence of a varying information quantity such as entropy.

5.6.3   Berry's Paradox

Consider a conventional (irreversible, in general) program P_{irr} that needs to be executed reversibly. A reversible forward execution $F_{rev}(P_{irr})$ of the irreversible program inevitably results in the generation of a memory trace M_{suf} that captures *sufficient* information to go back in the execution. Now, suppose we would like to minimize the length of the memory trace, to reduce the resources needed for the reversible execution. Essentially, we would like to know what would be the *necessary* information M_{nec} needed to go back in the execution. We know that the minimum necessary memory is bounded by the sufficient memory trace: $M_{nec} \leq M_{suf}$. We also know that the logged sufficient trace does contain all the information needed for reversal, perhaps with some redundant bits scattered inside the trace. Then, the question about the minimality of trace size for reversal can also be asked equivalently as follows: What is the shortest bit representation of M_{suf}? In particular, we can even allow the shorter representation to assume a computational component in its representation. In other words, the shortened representation could itself be a program that could be *executed* to regenerate the set of original bits encoded in the M_{suf} bits of the memory trace (see also Section 5.7). The program could, for example, encode the redundant bits into a fewer number of bits, and also detect and exploit patterns in the essential bits, and potentially perform other optimizations of such kinds. Consider a conventional (irreversible, in general) program P_{irr} that needs to be executed reversibly. A reversible forward execution $F_{rev}(P_{irr})$ of the irreversible program inevitably results in the generation of a memory trace M_{suf} that captures *sufficient* information to go back in the execution. Now suppose we would like to minimize the length of the memory trace, to reduce the resources needed for the reversible execution. Essentially, we would like to know what would be the *necessary* information M_{nec} needed to go back in the execution. We know that the minimum necessary memory is bounded by the sufficient memory trace: $M_{nec} \leq M_{suf}$. We also know that the logged sufficient trace does contain all the information needed for reversal, perhaps with some redundant bits scattered inside the trace. Then the question about the minimality of trace size for reversal can also be asked equivalently as follows: What is the shortest bit representation of M_{suf}? In particular, we can even allow the shorter representation to assume a computational component in its representation. In other words, the shortened representation could itself be a program that could be *executed* to regenerate the set of original bits encoded in the M_{suf} bits of the memory trace (see also Section 5.7). The program could, for example, encode the redundant bits into a fewer number of bits, and also detect and exploit patterns in the essential bits, and potentially perform other optimizations of such kinds.

Unfortunately, despite intuition about the potential for reductions in the number of bits, it is not possible to determine the shortest number of bits needed, as shown next by its relation to Berry's Paradox [Whitehead and Russell, 1925].

Berry's Paradox is best illustrated with the following example. The statement "the smallest number that cannot be defined in less than thirteen words" is a contradiction in itself because that number can in fact be defined by the preceding phase in quotes that is only twelve words long [Bennett, 1979]. An expansion of this basic conceptual approach includes the possibility of using any *program* to execute and output the desired bit representation (i.e., the set of words used in the definition of the number could in fact be a program that, when executed, gives the bit representation of the desired number). Berry's Paradox highlights an apparent contradiction of terms (or fundamental ambiguity and descriptive imprecision) in our attempt to determine the minimum bit length to completely define any number. Thus, the determination of the minimum length of a definition program becomes undecidable. The implication of Berry's Paradox to reversible execution is that we should not expect to answer the question of finding the minimal bit length encoding of (a program that recreates) the memory trace generated by forward execution to enable reversibility.

For example, suppose the forward execution of a program generates pseudorandom numbers using a high-quality generator. If we do not take into account the fact that a pseudorandom number generator underlies the behavior of the forward execution, the only recourse for reversibility is to record the sequence of random numbers to a memory trace. Suppose the trace is L bits long (which is the value of M_{suf} for this case). By definition of the randomness of the logged numbers, the trace cannot be compressed to much less than L bits, even with sophisticated compression techniques. However, suppose the size of the inverse generation *program* of the pseudorandom number generator is Q bits. Then, the trace can in fact be encoded by the following combination: (1) the Q-bit program for inverse generation, (2) w bits for the initial or final seed, and (3) $\log L$ bits to record the count of the number of invocations to the generator during forward execution. Thus, $Q + w + \log L$ bits are necessary and sufficient to reverse the execution (which is the value of M_{nec} for this case). This is in contrast to the larger number of bits (L) that would be logged to the memory trace without the knowledge of the pseudorandom number generation. This example provides an idea of the difficulty in determining the minimum number of bits that would be needed to augment any (irreversible) program in order to render it reversible.

5.7 Algorithmic Entropy

Another concept closely related to reversible computing is *algorithmic entropy*. Also known as *algorithmic complexity*, it is defined on a bit string as the length of the smallest program that, when executed on a universal computer, produces that bit string as output. Algorithmic entropy can be applied to the reversal of irreversible programs as follows. An irreversible program can be executed over a reversible deterministic machine by generating a history trace. After the trace is generated, we can attempt to compress the trace so that the number of trace bits is reduced. In general, we can consider replacing the history trace by a more condensed *program*, which when executed reproduces the original history bits. If the length of this encoded program happens to be independent of the forward execution length, we have a fixed energy cost because the cost to convert the history bits to clean ones is constant. Unfortunately, it turns out that the algorithmic entropy is not a computable function, and hence we cannot expect to develop any schemes to reduce the energy cost by relying on the ability to determine the algorithmic entropy of history bits. The formal definition of algorithmic entropy is described and its non-computable nature is proven next.

5.7.1 Definition

Let P be a program that, when executed on a universal computer U, results in output x. Let the bit length of the program P (since it is itself a sequence of bits) be the integer $|P|$. This length is inclusive of input, if any, needed by P to generate the output x. Let $|x|$ be the bit length of the output x. Then, P is an algorithmic encoding of x. If $|P^*| = \min |P|$ for all such P, then P^* is a minimal bit size encoding of x, and $|P^*|$ is the *algorithmic entropy* of x. This is also sometimes known as *Kolmogorov Complexity* [Kolmogorov, 1963, Shiryayev, 1993] or *Solomonoff Complexity* [Solomonoff, 1964a,b].

Trivially, $\alpha_1 + \log_2 x \leq |P^*| \leq \alpha_2 + |x|$, for some constants α_1 and α_2 that do not depend on x. This is because, for the lower bound, the program must embed within itself (in some encoding form) the number of bits in the output. For the upper bound, one can simply print the string x as it is, using an output statement of the universal computer. The constants α_1 and α_2 account for the simple infrastructure needed to represent the output operation.

5.7.2   Non-Computability

Let $C(x)$ be a function that computes the bit size of a minimal program P^* that can generate x as output on a universal computer. Then, the following result can be obtained:

Theorem 1 $C(x)$ *is not a computable function.*

Proof 1 *If $C(x)$ could be computed, let q be a program as shown in Algorithm 5.3. The program q prints the first integer x^* such that $C(x^*) =$*

Algorithm 5.3 Program q to disprove the computability of algorithmic entropy

1: **for** $x' = 0$ **to** ∞ **do**
2:    **if** $C(x') > |q|$ **then**
3:       **print** x'
4:       **halt**
5:    **end if**
6: **end for**

$|P^*| > |q|$. *However, because q also prints x^*, q can be seen as another program that generates x^*. Hence the same x^* can be encoded by q itself with fewer bits $|q|$ than $|P^*|$. This contradicts the minimality of P^* as claimed by $C(x^*)$. Thus, $C(x)$ cannot be computable.*

Corollary: One cannot write a program that can output the minimal program (thereby achieving the algorithmic entropy) needed to encode a given input bit pattern.

Note that the proof rests on the assumption that q is indeed an *algorithm*, that is, a procedure that provably halts. This requires the following two conditions to be met:

(a) For every $\tilde{x} > 0$, there exists an $\hat{x} > \tilde{x}$ such that $C(\hat{x}) > C(\tilde{x})$. This provides the guarantee that eventually an x' is found such that $C(x') > |q|$ as x' is increased from 0.

This seems intuitively satisfied, because otherwise, infinitely many numbers can be defined by a finite number of symbols. Interestingly, algorithmic entropy is also related to Berry's Paradox that deals with the minimum number of words to describe any given number. For example, the statement "The first number that cannot be described in less than thirteen words" can in fact be described by the preceding words in quotes (also, see Section 5.6.3). Berry's Paradox is essentially the same problem as the problem of determining the algorithmic entropy of a string. There exists a solution to Berry's Paradox [French, 1988] that gives a construction of an enumeration procedure for a finite set of words to define an infinite number of numbers; it uses the same expression (i.e., a constant set of symbols) that semantically describes any number larger than 1. This argument challenges the assumption that description sizes for numbers must asymptotically increase.

(b) $C(0) \leq |q|$. Because the definition of "zero" can be varied, we have to choose a system in which the algorithmic entropy of zero is not larger than $|q|$.

This condition seems trivial to satisfy because *we* can choose the language in which we describe zero, and we also have the choice of language for q. However, in light of the distinction between predetermined value *versus* computable value for representing any number [French, 1988], this may open a philosophical question, but we will not venture in that direction here.

The type of *diagonalization* argument technique used in the preceding proof underlies other related (equivalent) formulations such as with the "Busy Beaver" Turing machines [Michel, 2004, Harland, 2007]. This is also in line with the proof by McCarthy that inversion of programs is undecidable [McCarthy, 1956].

5.8   Further Reading

The independence of the concepts of thermodynamic reversibility and logical reversibility continues to be debated (see, for example, [Maroney, 2004, 2005]). Additional relations among thermodynamical concepts such as temperature and entropy have been applied to computer codes by H. Baker [Baker, 1992, 1994], with interesting analogies from physical systems containing reversible dynamics of billions of particles to computer systems containing reversible dynamics of billions of behavioral objects in computer memory. The operation of garbage collection sub-systems is shown to exhibit entropy increase and decrease with an increasing number of live objects that are allocated but potentially forgotten by the application program.

For a good overview of the Ehrenfest's Urn model, see Chapter 1.4 in the book [Kelly, 1979]. Additional treatment on Boltzmann's explanation to increase of entropy can be found in numerous works in the literature. Computer-based verification efforts [Orban and Bellemans, 1967, Levesque and Verlet, 1993] included simulations to verify the models and demonstrate that numerical effects can be discounted as a hidden source of such an entropy increase.

There have been, and continue to be, active and passionate debates on the reversible and irreversible aspects of the fundamental physical processes in nature. Many consider the microscopic laws as reversible, while some refute or qualify such a view. Despite being debated for centuries, the reversibility of classical mechanics is revisited even today for clarifications on the nature of reversibility of the simplest fundamental laws such as equations of motion expressed as Hamiltonian systems (e.g, [Hutchison, 1993, Roberts, 2012]). Besides such debates about microscopic laws themselves, there continue to be disagreements about the explanation offered in the form of macroscopic irreversible dynamics arising from reversible microscopic processes. For example, the arrow of time [Lebowitz, 1994] suggested by such an explanation is not universally accepted as the explanation for the arrow of time perceived in our daily life—objections are offered such as the gap left by the unanswered question about why the universe should be assumed to have started in a low entropy state in the first place.

The one-way evolution to an equilibrium state in systems such as the Ehrenfest's Urn model makes the notion of macroscopic irreversibility dependent on the quality of randomness introduced into the system, and the assumption that the outcome of the dice is independent of the state of the system is critically implicit without adequate justification when applied to reality. Perhaps the fairness or uniformity of randomness could be traced or mapped to the most fundamental sources of such uniformity, namely, the unbiased superposition of (pure) states at the quantum level; however, even that source of fairness is essentially an axiom, without which the explanation for one-way evolution (even if limited to the initial half of evolution before return to initial state) would become moot.

A critical point of relevance of all the aforementioned material to applications of reversible computing is the understanding that the presence of an arrow of time (or an increase of entropy) in a phenomenon does *not* imply that a computer-based reversible simulation of that phenomenon would necessarily accumulate memory during simulation.

Chapter 6

Theoretical Computing Models

6.1 Overview

According to the widely accepted Church–Turing thesis, every computable function (or algorithm) can be expressed as a Turing Machine program [Ben-Amram, 2005]. Hence, to theoretically examine the reversibility of any computation in general, it is sufficient to focus on the reversibility of computations realized by Turing Machine programs. If there were a way to execute any Turing Machine program reversibly, then any computable function can be executed reversibly. With these considerations, the classical Turing Machine model is examined here and shown to be *irreversible*. Next, the construction of a Reversible Turing Machine model is described such that the machine is not only *reversible* by design, but also capable of reversibly simulating any ordinary Turing Machine program, thus proving that any computable function can be reversibly executed.

With regard to the relation between reversible circuits on the one hand and reversible universal machines on the other, a fundamental difference makes them "decisively different" [Li and Vitanyi, 1996]. While it is possible to design custom circuits on a problem-specific basis, universal machines must be designed to accept any program as input and reversibly execute them. This difference results in different design criteria and consequently different spatial and temporal characteristics and constraints.

6.2   Turing Machine Model

While there are many variants of the basic Turing Machine model, all of them can be mapped to a basic form, namely a canonical Turing Machine that operates with one infinite tape and one head. All variants are computationally equivalent in the sense that only a polynomial cost factor is incurred in space and/or time when realizing the variant model on the canonical model. Thus, it is sufficient to consider the reversibility of a 1-tape, 1-head canonical machine.

Without loss of generality, we will use the definition of a canonical Turing Machine as a closed system defined in Table 6.1. This traditional model can be used to represent and realize any computable function. While much theory exists on Turing Machines in the context of forward execution, it is known that this model is not reversible because the sequence of transitions performed on this model cannot be retraced uniquely. Also, it has not been clear how to define a *reversible* Turing Machine, and if it is indeed possible to achieve reversible execution in general for any computable function. Following a series of efforts by several researchers to resolve this problem, this matter was finally settled in a seminal paper in 1973 by Charles H. Bennett, who showed that it is possible to not only define a reversible execution model, but also simulate the execution of any conventional (forward-only) Turing Machine program on the new reversible model [Bennett, 1973].

The solution to the generalized reversibility problem involves three important aspects: (1) definition of new semantics of an appropriate reversible machine, (2) an algorithm for mapping any conventional Turing Machine program to the new reversible machine, and (3) analysis and minimization of space and time costs of the reversible machine when simulating conventional Turing Machine programs. The schemes for each of these aspects developed in [Bennett, 1973] are described next.

TABLE 6.1: Definition of a Canonical Irreversible Turing Machine

A canonical Turing Machine M is a quintuple $\langle Q, q_0, F, S, \delta \rangle$, where:

Q	=	Set of possible states $\{q_0, \ldots, q_{f-1}\}$ for the control unit
q_0	=	Initial state of the control unit
$F \subset Q$	=	Set of final states of the control unit that halt the machine
S	=	Set of tape symbols $\{s_0, \ldots, s_{z-1}\}$, $s_0 = b$ being the blank symbol with which all slots except for the input slots of the tape are initially filled
δ	=	Transition function of the control unit

$$\delta : Q \times S \to Q \times S \times D,$$

where $D = \{L, R\}$ denotes left or right movement of the tape head, and the function is defined by a set of transition rules $\{L_0 \ldots L_{N-1}\}$, each of the form

$$L_i : (q, s) \to \langle q', s', m \rangle,$$

denoting that when the control unit is in state q and the current tape symbol is s, the symbol s is overwritten with the symbol s', the state is changed to q' and the tape head is moved either left ($m = L$) or right ($m = R$).

The machine has f states, z symbols, and N rules.

6.3 Sources of Irreversibility in the Turing Machine Model

There are two fundamental sources of irreversibility in the conventional Turing Machine model. In Bennett's words, the machine will be reversible if and only if the δ function has non-overlapping ranges. The overlap arises from two aspects of the transition rules:

1. **Write-and-move semantics**: The semantics of the read-write-and-move operation of each transition rule introduces ambiguity when retracing the forward execution path. The Turing Machine head, upon reading the tape symbol s at the slot of current head location $T[i]$, can overwrite the symbol with a new symbol s' and then move the head to the left slot $T[i-1]$ or to the right slot $T[i+1]$ from the current slot. During reverse execution, it would be unclear whether a current slot $T[j]$ was reached from the forward execution by moving right from the left slot $T[j-1]$ or by moving left from the right slot $T[j+1]$. In general, it

would be impossible to disambiguate between the two possibilities, and such ambiguities can occur at any point in execution.

2. **Many-to-one transition state mapping**: The state transition function may, in general, include two or more transition rules, say,

$$(q_i, s_i) \rightarrow \langle q_k, s_k, m_i \rangle$$

and

$$(q_j, s_j) \rightarrow \langle q_k, s_k, m_j \rangle$$

that result in the same next state $(q_k, .)$. During backward execution, the correct previous state $(q_i$ or $q_j)$ may not be recoverable from the current state q_k because of the ambiguity about the previous state from which the current state was reached during forward execution. In other words, if the forward state transition function is a *many-to-one* mapping, then the machine is irreversible due to the transition function.

Note that the machine may be *deterministic* (i.e., the transition function is not a *one-to-many* mapping), and yet it could be *irreversible*. Determinism in the forward execution is different from reversibility. If the reverse state transition function is deterministic (i.e., the function is a *one-to-one* mapping), reversibility of the forward function is ensured. Also note that, in general, the forward function need not be an *onto* function; that is, not every state is reachable from the initial state(s). For example, the initial states need not appear as the target states for any intermediate states.

6.4   Definition of a Reversible Model

The two sources of irreversibility were overcome by Bennett using the following steps:

- A new spontaneous transition possibility is introduced by adding a special *solidus* symbol (/), which acts as a wild card, matching any symbol at the current tape head; alternatively, the action of solidus can be viewed as not reading the tape or ignoring the current symbol.

- The form of transition rules is redefined to split the combined write and move operations into two separate rules, one for writing and the other for moving the head.

- The machine formalism is generalized to accommodate multiple tapes instead of the single tape of the conventional model.

■ The transition rules are relaxed to accommodate multi-head operation, by changing the symbols and tape movement specifications into
  k-element vectors (for k tapes) instead of scalars for 1-tape operation.

■ The machine is augmented with two additional tapes: the original *working tape*, a new tape called the *history tape*, and another called the *output
  tape*.

6.4.1 Rewriting Transition Rules: Quintuples to Quadruples

In order to remedy the first basic source of irreversibility in the conventional
Turing Machine model, Bennett defined a variant of the write-and-move semantics by separating the write operation from the move operation. Every
write-and-move rule of the conventional Turing Machine is converted into two
separate rules: one rule that writes a symbol and transitions to a special intermediate state unique to that transition, and the second rule that moves
the tape head left or right while transitioning from the special intermediate
state to the target state of the original rule. Thus, a conventional quintuple
transition rule of the form

$$L : (q, s) \rightarrow \langle q', s', m \rangle$$

is split into two quadruples,

$$L : (q, s) \rightarrow (\overline{q'}, s')$$

and

$$\overline{L} : (\overline{q'}, /) \rightarrow (q', m).$$

The use of the solidus (/) in place of a symbol indicates that the head of
the tape does not read any symbol, that is, the transition is spontaneous and
independent of the current symbol. A new state $\overline{q'}$ unique to each original rule
is added to the set of states Q. The first part writes the symbol and the second
part spontaneously moves the head left or right. The number of rules is thus
increased by a factor of 2, as every conventional quintuple rule is translated
into two rules in the new quadruple form.

6.4.2 Adding History and Output Tapes

After the transition function of the conventional program is rewritten in the
new form, the irreversibility of the write-and-move semantics arising out of the
non-commutativity of the write and move operations is eliminated. However,
the second (which is the more challenging) source of the irreversibility problem
still remains to be resolved, namely the overlap of ranges. In general, there is no
easy way to avoid the overlap without introducing additional storage. The only
general-purpose solution for reversibility is to maintain a log of the sequence of

transitions, which can be retraced for reversal of the forward sequence. Upon completion of the forward execution, the log is consulted to bring the machine back to the original starting state. However, simply reversing the execution also destroys the computed output string as well. Hence, to make the program reversible, two sub-problems need to be solved: how to keep a log of actions and how to save a copy of the final output.

These are addressed by introducing two additional tapes to the machine, with one head per tape. The original tape is called the *working tape*, and the added tapes are called the *history tape* and the *output tape*, respectively. The basic reversible computation approach proceeds as follows. The original (irreversible) program is allowed to operate normally, reading from and writing to the working tape. However, before every transition of the original program, the identity of the transition rule that is being executed is recorded in a sequence on the history tape. When the original program enters a halting state, the new program copies the output string, currently found on the working tape, onto the output tape. Then, a reverse mode is initiated whereby a set of new rules that are disjoint from the original (forward execution) rules is used to undo the action of every forward transition in reverse order. Because the history tape holds a record of the identity of every forward transition that was made, no ambiguity arises as to which transition is to be undone in sequence. Moreover, as part of the reversal of each transition, the history tape is also cleaned (reversibly) by resetting the most recently consulted transition number to the blank symbol. When the end of the trace on the history tape is reached, it is evident that the entire execution has been undone. At this point, the working tape would have recreated the original input, the history tape would have been reset to all blanks, and the output tape would contain a copy of the output computed by the original machine.

To enable this 3-tape operation, the original 1-tape formalism is enhanced to accommodate 3-tape transition rules. This is achieved by relaxing the transition rule to be triggered by a vector of k symbols (in this case, $k = 3$) corresponding to one symbol per tape, instead of action on a single scalar symbol of the 1-tape machine rule. Similarly, the action of each rule is relaxed by specifying a vector of individual actions on each tape. Note that the trigger for each tape may include a symbol of that tape's alphabet or the solidus / that acts as a wild card. Similarly, the action on each tape may be to either move the head or overwrite the current symbol with a symbol of that tape's alphabet.

6.4.3 Canonical Turing Machine Model

In the development of the Reversible Turing Machine, the canonical form of the irreversible Turing Machine is replaced by an equivalent form that is easier to treat for reversibility. This form, called the *standard* form by Bennett [Bennett, 1973], is the one that will be rendered reversible in the following sections.

TABLE 6.2: Definition of a Standard Irreversible Turing Machine

Given a canonical Turing Machine $M = \langle Q, q_0, F, S, \delta \rangle$ in the form of Table 6.1, a *Standard* Turing Machine M_S is obtained as follows:

δ	is modified by adding one unique intermediate state per rule and splitting every write-and-move rule into two rules, one to write and another to move
Q	is increased to include all the introduced unique intermediate states of the modified rules

The standard Turing Machine thus has twice as many transition rules as its canonical version.

For convenience, the following additional assumptions are made in the standard form of the irreversible Turing Machine as shown in Table 6.2:

■ The input and output are assumed to be arranged such that there are no blank symbols embedded within either the input or the output strings.

■ The tape head is initially positioned immediately left of the first input symbol on the tape.

6.5 Mapping Conventional Model Programs to a Reversible Model

The reversible execution is achieved in a three-stage algorithm. In the first stage, the normal computation proceeds, but the identity of every transition is logged to the history tape. In the second stage the output generated at the end of the first stage is copied onto the output tape. Finally, in the third stage, the entire computation is undone, using the working and history tapes. In order to realize these three stages, the original program must be enhanced with additional states, additional alphabets/symbols, and new transition rules. The enhanced machine that realizes the three stages is shown in Table 6.3.

Given a Standard (irreversible) Turing Machine M, the transition rules for the Standard Reversible Turing Machine M_R are obtained as follows for the three stages:

Stage 1 Each quadruple transition rule L_i from state (y, s) to (y', s') is split into two rules, L_{iw} and L_{ih}. The rule L_{iw} performs the same action on the working tape as L_i does, but instead of transitioning to q', transitions to a unique intermediate state r_i. The rule L_{ih} picks up from r_i, records the integer symbol i (to log the identity of the transition rule) on the history tape and then transitions from $\bar{q}_i$ to q'. The

TABLE 6.3: Definition of a Standard Reversible Turing Machine

Given a standard Turing Machine $M = \langle Q, q_0, F, S, \delta \rangle$ in the form of Table 6.2, a *Standard Reversible* Turing Machine $M_R = \langle Q_R, S_R, \delta_R \rangle$ is a 3-tape closed system, where:

Q_R	$=$	$Q \cup \{\bar{q}_0, \ldots, \bar{q}_{f-1}\} \cup \{r_0, \ldots, r_{N-1}\} \cup \{\bar{r}_0, \ldots, \bar{r}_{N-1}\} \cup$ $\{a_0, a_1, \bar{a}_0, \bar{a}_1\}$ are the states for the control unit, such that $q_0 \in Q$ is the initial state, and $\bar{q}_0 \in Q$ is the only final halting state of the control unit
S_R^W	$=$	S is the alphabet for the first (working) tape
S_R^H	$=$	$\{0, \ldots, N-1, b\}$ is the alphabet for the second (history) tape
S_R^O	$=$	S is the alphabet for the third (output) tape
$S_{R/}$	$=$	$\left\{ x \middle| x = \begin{bmatrix} s_w \in S_R^W \cup \lambda \\ s_h \in S_R^H \cup \lambda \\ s_o \in S_R^O \cup \lambda \end{bmatrix} \right\}$ is the set of vectors of tape symbols or spontaneous actions that trigger a transition
S_{RD}	$=$	$\left\{ x \middle| x = \begin{bmatrix} s_w \in S_R^W \cup D \\ s_h \in S_R^H \cup D \\ s_o \in S_R^O \cup D \end{bmatrix} \right\}$ is the set of vectors of actions of a transition, either overwriting with symbols or moving the head, specified on a per-tape basis
λ	$=$	$\{/\}$ denotes a wild card match (or ignorance) of the current symbol
$D =$	$=$	$\{L, R\}$ denotes left or right movement of the tape head
S_R	$=$	$S_{R/} \cup S_{RD}$ is the combined set of vectors defining the triggers and actions of the machine
δ_R	$=$	Transition function of the control unit

$$\delta_R : Q_R \times S_{R/} \to Q_R \times S_{RD},$$

defined by the set of transition rules $\{L_0 \ldots L_{N_R-1}\}$, where $N_R = 4N + 2z + 3$. Each transition rule is of the form

$$L_i : (q, s) \to (q', s'),$$

indicating that the rule is triggered when the control unit is in state q and the current tape symbol vector is s. For each tape k, if $s'[k] \in D$, the tape head is moved; otherwise, the symbol on tape k is overwritten with $s'[k]$. The state is then changed to q'.

The machine has $2f + 2N + 4$ states and $4N + 2z + 3$ rules. Further, it uses tape alphabets containing z, $N + 1$, and z letters, for the working, history, and output tapes, respectively.

semantics of the original machine are thus retained unchanged with respect to the states as well as the values on the working tape, but a side effect is introduced, namely a recording of the transition rule number on the history tape. Because there are f original states and N original rules, and one new state is introduced for each original rule and two new rules for every single original rule, this stage defines $f + N$ states and uses $2N$ rules.

Stage 2 From the halt state q_{f-1}, the output is copied from the working tape to the output tape. This is done in three parts: (1) two rules are used to merge the control from state q_f of Stage 1 to state a_0 of Stage 2, (2) $2z$ rules are used to transfer the output symbols from the working tape to the output tape, and (3) one rule is used to transfer control from the final state $\bar{a}_1$ of Stage 2 to the initial state $\bar{q}_{f-1}$ of Stage 3. Overall, Stage 2 defines 4 new states and uses $2z + 3$ rules.

Stage 3 From $\bar{q}_{f-1}$, the original rules of Stage 1 are undone in reverse order while cleaning up the history tape symbols by overwriting the transition number symbols with blanks, one at a time in the reverse order. Similar to Stage 1, this stage defines $f + N$ states and uses $2N$ rules.

The full specification of the standard Reversible Turing Machine is given in Table 6.3.

6.6 Universality of Computation and Its Reversal

The elegance of the transformation algorithm is made evident from the fact that the algorithm applies to any Turing Machine. In particular, this implies that the transformation is equally applicable to any *Universal* Turing Machine (UTM) as well. A UTM is a Turing Machine that takes any arbitrary Turing Machine program P specified as input to the UTM (along with the input I needed for that input program P), and simulates the operation of P over I, generating the output O. Because any UTM is itself a Turing Machine, any UTM can also be rendered reversible by taking the program P_{UTM} and generating a reversible version of that UTM using the aforementioned conversion algorithm. Thus, the existence of reversible Universal Turing Machines is evident. Furthermore, any computable function can be executed reversibly by any general-purpose reversible Universal Turing Machine.

6.7   Space and Time Complexity of Reversible Execution

We will now examine the space and time complexities of the simple reversible simulation and examine two additional schemes that improve the space and time. It turns out that the simple version is in fact optimal with respect to time but vastly suboptimal with respect to space. The two variants of this scheme improve the space complexity dramatically at the cost of an increase in computational time.

In what follows we will omit implementation-specific constant multipliers and constant additives in complexity; for example, the time complexity of T will be understood as $\Theta(T)$.

6.7.1   Complexity of Simple Reversal with One Segment

Although the preceding organization of the Reversible Turing Machine computes the output reversibly and cleanly with only the output left at the end of the computation, there is one major drawback, namely the length of the history tape used at runtime. The size of temporary storage required by the machine on the history tape is proportional to the total length of execution. If T is the length of execution in terms of the number of transition rules executed by the machine, then the required length of the history tape is equal to T. Thus,

$$Time(T) = T$$

and

$$Space(T) = T.$$

Because T can be as large as 2^S because any non-looping computation on space of size S can iterate through all possible states of $\Im$, the space complexity with this scheme can be very large.

6.7.2   Partitioning Execution into Two Segments

Bennett proposed an approach by which the tape size can be reduced from T, but at some extra computation cost. In this approach to reducing the history tape length, the total execution length (number of computational steps) is logically partitioned into two halves called segments. Soon after the execution of the first half segment is completed, the machine is stopped and its current configuration C is saved on a vacant region of the output tape. In C, the following information is saved:

1. A snapshot of the current string C_W on the working tape and the tape head position,

2. A snapshot of the current string C_O on the output tape and the tape head position, and

3. A copy of the current control state q.

Thus, C contains all the information needed to re-initialize the machine and restart it at a later time from the midpoint. The machine is now executed in reverse, thereby cleaning out the working tape and the history tape. Note that the length of the history tape used so far is only $\frac{T}{2}$, and the amount of computation is $\frac{T}{2}$ for forward execution of the first segment and $\frac{T}{2}$ for reverse execution of the first segment. The machine is then restarted from the mid point using the saved configuration C, and executed to the end. The output of the computation is saved on the output tape, and the execution is reversed to clean up the history tape. At this point, however, the machine is still left in the configuration represented by C. For the overall execution to be reversible, the machine should be restored to the initial state, in which the working tape is left with only the original input, and the history tape is empty. However, the working tape is left with the values from C, which need to be erased. Two key observations are made here:

1. Blind erasure is an irreversible operation. In other words, unconditionally overwriting all the non-blank symbols of the working tape with blanks is irreversible, and hence the working tape cannot be cleaned of C in this fashion.

2. Given another copy of C, the previous copy of C can be reversibly erased, because the copy operation can be realized using the exclusive-or $\oplus$ bit operation semantics, which is reversible (in fact, a self-inverse). If b represents the initial blanks on the tape, and $b' = C$ is the copy on the tape, then the following holds:

$$b' \leftarrow b \oplus x \text{ (Copies } x \text{ into blank } b) \text{ and}$$

$$b \leftarrow b' \oplus x \text{ (Erases } x \text{ from } b').$$

If we are able to *reversibly reconstruct* a copy of C, then this new reconstructed copy of C can be used to erase the old copy of C. Reversible reconstruction of C can be achieved by re-executing the first half segment forward. Thus, the first segment is re-executed, C is erased, and the first segment is undone by executing in reverse. This entire operation is illustrated in Figure 6.1.

Using this segmentation approach, the required size of the history tape can be reduced by half, at the cost of some extra computation. Let $Time(T)$ be the time taken to execute the program using no segmentation, and $Space(T)$ be the memory consumed using no segmentation. Let $Time(\frac{T}{2})$ be the time taken using the 2-segment approach, and $Space(\frac{T}{2})$ be the space consumed in the 2-segment approach. We know that $Time(T) = 2T$ because of one full sweep of execution in the forward direction, followed by one full sweep in the reverse direction, and $Space(T) = \alpha$ for some α independent of T. Also, $Time(\frac{T}{2}) = 1.5Time(T)$, because, with 3 parts of forward execution and 3 parts of reverse execution, each of length $\frac{T}{2}$, the time for the 2-segment

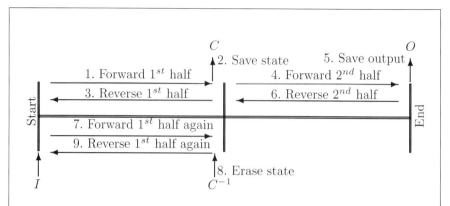

1. Forward execution from initial state with input I to midpoint
2. Saving the half-way state C
3. Reverse execution from midpoint back to initial state
4. Forward execution from midpoint to final state with output O
5. Saving the final output O
6. Reverse execution from final state back to midpoint
7. Forward re-execution from initial state with input I to midpoint
8. Reversibly erasing C with C^{-1}
9. Reverse execution from midpoint back to initial state

FIGURE 6.1: Reversible execution with half trace size.

execution is $Time(\frac{T}{2}) = 6\frac{T}{2} = 1.5Time(T)$. At this point, the savings in history tape size is only a constant factor from the original length. However, we can apply the same approach on each of the half-segments themselves by recursion. Thus, the total time $Time(T)$ can be expressed as the relation

$$Time(T) = 6Time(\tfrac{T}{2})$$

and

$$Time(1) = 1,$$

giving

$$Time(T) = 6^{\log_2 T} = T^{\log_2 6} = T^{1+\log_2 3} \approx T^{2.59}.$$

Let $S = Size(C)$, which is the total size of the machine state. The space requirements can be expressed by the relation

$$Space(T) \leq S + Space(\tfrac{T}{2})$$

and

$$Space(1) = 0,$$

giving

$$Space(T) \leq S\log_2 T \leq S\log_2 2^S = S^2,$$

because the computation time T on state C is bounded by the number of possible state values of C, which is 2^S. Note that the same memory space used by the first segment will also be reusable by the second segment. Hence, in the preceding equation for $Space(T)$, the term $Space(\frac{T}{2})$ is *not* multiplied by 2.

Thus, any computation can be reversibly computed using the Reversible Turing Machine in $T^{1+\log_2 3}$ time using space S^2. This method is also sometimes called the $\log 3$ method, due to the additional time factor of $T^{\log_2 3}$ relative to the simple scheme that only takes linear time T. The reduction in space requirement is significant because, without the segmentation approach, although the time is T, the space required would be $S + T$ which can be as large as $S + 2^S$.

6.7.3 Partitioning Execution into g Segments

The time and space complexity can be further improved by generalizing the two-segment approach to operate with $g \geq 2$ segments. This is illustrated in Figure 6.2. The time complexity is given by the relation

$$Time(T) = (4g - 2)Time(\tfrac{T}{g})$$

and

$$Time(1) = 1,$$

giving

$$Time(T) \leq (4g)^{\log_g T} = T^{\log_g 4g}.$$

Also, the space requirements can be expressed by the relation

$$Space(T) \leq (g - 1) \cdot S + Space(\tfrac{T}{g})$$

and

$$Space(1) = 0,$$

giving

$$Space(T) \leq (g - 1) \cdot S \cdot \log_g T.$$

6.7.4 Optimizing g for Minimal Total Space

With g segments, the total space used is the reduced history tape size $\frac{T}{g}$ used per segment (which is reusable across segments) added to the output tape space to store $g - 1$ snapshots of C generated at the end of each segment. Thus, the total space Z, given by $Z = \dfrac{T}{g} + (g-1)S$, is minimized with respect

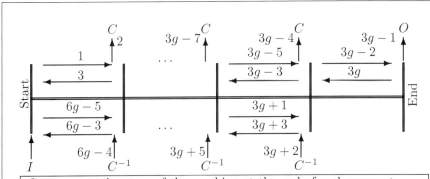

C represents the state of the machine at the end of each segment
C^{-1} represents the erasure of C corresponding to that segment

FIGURE 6.2: Reversible execution with g-fold smaller trace size.

to g by setting the first derivative of Z to zero (the second derivative being positive), giving $g^* = \sqrt{\frac{T}{S}}$ as the optimal value of g for minimal total memory. At this minimal value, the information needed for reversal gets divided into two equal halves of size $\sqrt{TS}$ each. One half is stored on the history tape and the other half appears in the form of snapshots written to the output tape. This is in line with the intuitive expectation that the optimal memory use reflects symmetry between time and space and also between history tape and output tape. Thus, at the optimal value of $g = g^*$, the total optimal space, Z^*, is given by

$$Space(T) = Z^* = 2\sqrt{TS}.$$

Recall that the simple unsegmented scheme incurs a space cost that is linear with respect to total execution time T. Relative to that, the optimized cost using the segmented scheme reduces the space cost by a factor $\alpha = \frac{T}{Z^*} = \frac{1}{2}\sqrt{\frac{T}{S}}$.

6.7.5 Generalizing the Time-Space Trade-Off

For execution length T and tape snapshot size S:

- ■ The simple scheme takes $Time(T) = T$ and $Space(T) = T$,

- ■ The 2-segment scheme takes $Time(T) = T^{1+\log_2 3}$ and $Space(T) = S \log_2 T \leq S^2$, and

- ■ The g-segment scheme takes $Time(T) = T^{\log_g (4g-2)}$ and $Space(T) = (g-1)S \log_g T$.

Examining these relations, it is clear that one gains either in time or space, but the schemes do not provide an effective continuum of solutions in the spectrum of time-space trade-off.

Incidentally, the time-space trade-off is similar to the same trade-off that arises in automatically converting linear recursive programs into iterative programs without recursion. It turns out that the problem of evaluating the time-space trade-off for recursive programs was analyzed by Chandra [1972], preceding the technique of Bennett [1973], and before the trade-off algorithm of Bennett [1988]. The latter provides a lever by which the reversible time can be reduced to come arbitrarily close to the irreversible time, but at an extremely large cost in space.

The inordinate space cost was uncovered by Levine and Sherman [1990] in some corrections to the time and space complexities of Bennett [1988]. The actual space complexity of Bennett's algorithm is given as

$$Space(T) = S + S \log \frac{T}{S}.$$

This shows that the space is in fact larger than that of the simplistic analysis by $S \log \frac{T}{S}$, which disappears if $T \approx S$ (i.e., if computation is linear in space), but can be large when $T = 2^S$. The bounds on the time and space were tightened even further as

$$Time(T) = \frac{T^{1+\epsilon}}{S^\epsilon}, \text{ and}$$

$$Space(T) = \epsilon 2^{1/\epsilon}(S + S \log \frac{T}{S}), \text{ where}$$

$$g = k^n, \epsilon = \log_k (2k - 1), n \geq 1, \text{ and } k \geq 1.$$

In particular, the value of k can be increased as a control parameter of the execution to deliver different time and space trade-offs (using the Bennett [1988] scheme). Relative to the simplistic analysis, the space using this more detailed analysis is larger by a factor equal to $\epsilon 2^{1/\epsilon}$. This factor represents a dramatic increase in space needs as $\epsilon \to 0$, if ϵ is desired to be reduced to make the reversible time approach the original irreversible time. Thus, the trade-off appears as one between the exponent value in reversible time and the factor value in reversible space. This limits the asymptotic complexity of the scheme, but implications to actual implementations are unclear until the schemes are attempted in actual reversible computer designs.

To arrive at an even more completely generalized set of strategies that deliver varied time and space complexities for reversible computation, an ingenious abstraction called the *Pebble Game* has been developed, as described next, solutions to whom directly correspond to space-time strategies for Reversible Turing Machines.

6.8   Pebble Games

The Pebble Game, originally mentioned by Charles Bennett [Bennett, 1988] and later explored in greater detail by others, consists of the following elements, as shown in Figure 6.3.

- An array of places (also called nodes) on a linear board are contiguously numbered $0 \ldots G$. Each place can accommodate one pebble.

- Initially, a pebble is already placed in the place numbered 0 (leftmost node in the array).

- A free pool of N pebbles is available aside from the board.

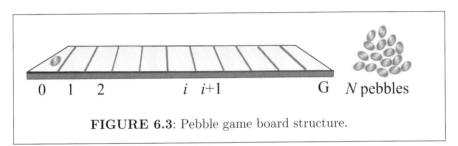

FIGURE 6.3: Pebble game board structure.

6.8.1   Rules and Objective

The single-player game is played as follows, with two rules and one objective. Different optimization goals may be defined with varying optimality criteria, resulting in different optimal strategies to meet the objectives.

- **Rule 1** A pebble from the free pool can be placed in an empty place $i + 1$ if there is already a pebble in place i (see Figure 6.4).

- **Rule 2** A pebble from place $i + 1$ can be removed from the board and returned to the free pebble pool only if place i has a pebble (see Figure 6.5).

- **Game Ending** Starting with an empty board except for the initial pebble at place 0, and conforming to the two preceding rules, place a pebble at the last place, G.

- **Optimization** Find the minimum number of pebbles in the initial free pool necessary to place a pebble in place G. Or, find the minimum number of placement and removal steps needed to place a pebble in place G. In general, find the relation between the number of steps and the number of free pool pebbles needed to satisfy the objective.

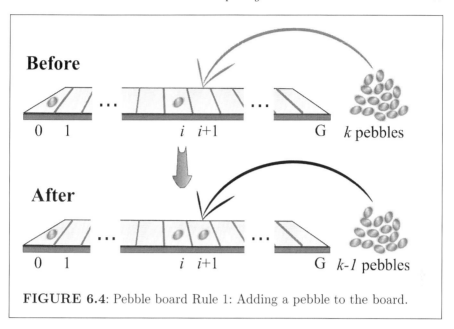

FIGURE 6.4: Pebble board Rule 1: Adding a pebble to the board.

6.8.2 Analogies with the Reversible Turing Model

In relation to simulation of irreversible computations on reversible machines, the operations in the pebble game have the following analogies:

- The pebble at place 0 at the beginning corresponds to the original input to the machine.

- The objective of placing the pebble in the last place G corresponds to the arrival of output from the machine.

- Placement of a free pebble on an empty place by Rule 1 is analogous to saving a copy of the current state of the machine.

- Returning a pebble from the board to the free pool by Rule 2 is analogous to reversibly erasing a saved state of the machine.

- The maximum number of pebbles present on the board at any time during the game is the same as the space complexity of the reversible simulation.

- Every strategy that meets the objective of the game provides a corresponding solution to the problem of reversible simulation of irreversible machines. In particular, extremes such as optimization for the minimum number of free pebbles needed at the beginning, or the minimum number of steps to reach the end, map to minimum space and minimum time, respectively, for the machine simulation.

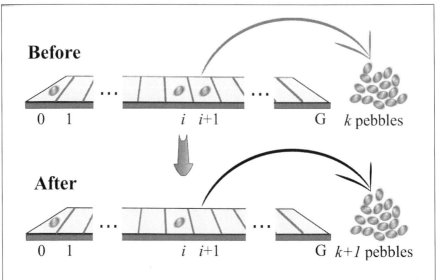

FIGURE 6.5: Pebble board Rule 2: Removing a pebble from the board.

■ At the end of the game, if there are any pebbles left on the board other than in places 0 and G, they correspond to the willingness of the user of the computation to accept the leftover pebbles as part of useful output. If not, the objective can be redefined to require the end state to contain pebbles only in the end places 0 and G.

6.8.3   Complexity Analysis

To apply the pebble game to reversible execution, we use $G = T$, where T is the irreversible execution time and G is the number of places on the board. When $N = 0 + O(1)$ pebbles are used, the game gives the exponential time method; and when $N = \log T + O(1)$ pebbles are used, the two-segment method is obtained [Buhrman et al., 2001]. In general, with $N + O(1)$ pebbles, the game gives the complexities as

$$Time(T) = 3^N \cdot S \cdot 2^{O(T/2^N)}$$

and

$$Space(T) = O(NS).$$

6.8.4   Partial Reversibility

Additionally, if the removal of any pebble from the board is in violation of Rule 2, such removal is analogous to irreversible erasure of a saved state in the simulation. Such an invalid move of the pebble (or irreversible erasure of a saved state) provides for a relaxation of the original problem of pure reversibility by introducing some irreversible operation. While purely reversible execution offers the schemes and trade-offs as presented previously, slight relaxations in the problem can potentially provide more flexibility, and consequently more efficiency, with respect to space and time. One such relaxation is the allowance for some irreversibility [Li and Vitanyi, 1996, Li et al., 1998], which can occur if one allowed for some irreversible erasures during the execution of the machines. In the context of pebble games, the equivalent of an irreversible erasure occurs whenever a pebble at place $i + 1$ is removed in the absence of a pebble at place i. This relaxation results in interesting trade-off between space, time and the number of irreversible operations [Li and Vitanyi, 1996]. Possibilities of limited irreversibility relaxations have been identified in special computing technologies such as quantum computing in which any measurement becomes an irreversible erasure. Strategies for cleverly deciding on the number and precise moments of making such irreversible operations are possible to develop using the relaxed version of the pebble game.

6.9   Further Reading

The technique of computing an irreversible machine as part of an enclosing reversible machine was presented in Lecerf [1963], not with considerations of thermodynamics, but for analyzing decidability properties of reversible transformations. Moreover, Bennett's technique of copying the computed answer before initiating the reversal was not part of this earlier work by Lecerf.

Time and memory trade-offs for energy savings are further analyzed by Li and Vitanyi [1996] and Li, Tromp, and Vitanyi [1998]. Additional refinements in time and space bounds are presented by Buhrman, Tromp, and Vitanyi [2001].

Another Reversible Turing Machine is defined in [Axelsen and Glück, 2011] from first principles, without being constrained to simulate the functioning of the older (forward-only) Turing Machine model. A reversible abstract machine is proposed in [Lienhardt et al., 2012] for rendering concurrent programs reversible.

The reversible machine algorithms discussed here assumed that the total length T of execution is known in advance. While such an assumption is reasonable for computations in which the total runtime is known (as in the case with sorting algorithms, for example), it needs to be refined for cases in which the runtime is not known a priori. An amended meta-algorithm for unknown computing time that works on top of known-T algorithms was presented by Li, Tromp, and Vitanyi [1998].

Chapter 7

Relaxing Forward–Only Execution into Reversible Execution

7.1 Overview of Paradigms

The traditional notion of forward–only execution seems to be comprehensible and acceptable quite naturally. However, when the notion of reversibility is introduced, it is not entirely intuitive as to how the forward–only notion may be relaxed to accommodate reversibility. Although many find the notion of computation easy to understand as a function $y = P(x)$ that is applied on input bits x to obtain output bits y, a simple natural notion of how this view can be altered for reversible execution is not clear, *in general*. Nevertheless, based on research into using reversible computation in various contexts over the past few decades, multiple relaxations have taken shape. Among the relaxations, originally motivated by finding energy-optimal execution for reversible computing, the paradigm of Compute–Copy–Uncompute came into vogue (described later in this chapter). The application of reversible execution to synchronization problems arising from on-chip speculative execution and inter-processor virtual time synchronization resulted in the development of another paradigm, namely, Forward–Reverse–Commit. The use of a notion of reversibility in popular user-oriented applications gave rise to yet another paradigm, namely Undo–Redo–Do. These three paradigms are described next. Additional paradigms are possible, which may be discovered and developed in the future.e

7.2 Compute–Copy–Uncompute Paradigm

In the most well-known use of reversible computing, energy loss is reduced by avoiding blind erasure of bits. This is conceptually achieved using the famous theoretical algorithm by Charles H. Bennett [Bennett, 1973] (and its later variants) whose essential operation is: compute, copy the output, and then uncompute (sometimes called the "Bennett trick" in the literature). This is the default approach that is commonly associated with classical approach to reversible computing. Given a program P, a traditional computing system executes P in forward-only mode, $\overline{F}(P)$, and generates the desired output. A reversible computing system, utilizing a modified forward function $F(P)$ and its inverse $R(F(P))$, first executes $F(P)$, performs a copy operation $Y(F(P))$ to save the desired output, and then invokes the inverse $R(F(P))$ to clean up the effects of $F(P)$. This is illustrated in Table 7.1, and a schematic view of the control flow across all components of this paradigm is shown in Figure 7.2.

TABLE 7.1: Relaxation of Forward–Only Execution into the Compute–Copy–Uncompute Paradigm

Forward–only	Compute–Copy–Uncompute Execution
$\overline{F}(P)$	$CCU(P) \equiv F(P) \rightsquigarrow Y(F(P)) \rightsquigarrow R(F(P))$

Notation		
P	$=$	Program code fragment
$\overline{F}(P)$	$=$	Traditional forward-only execution of P
$F(P)$	$=$	Reversible forward execution of P
$Y(F(P))$	$=$	Saving a copy of output from $F(P)$
$R(F(P))$	$=$	Reverse execution of P after $F(P)$
$X \rightsquigarrow Y$	$=$	X followed by Y

7.2.1 Equivalence Conditions

Clearly, the forward–only execution and the Compute–Copy–Uncompute execution are considered equivalent if the output from $\overline{F}(P)$ is contained in the copy of the output made by $Y(F(P))$.

7.3 Forward–Reverse–Commit Paradigm

The Forward–Reverse Commit paradigm is very useful in situations where a program fragment can be executed "ahead of time," but is reversed if the

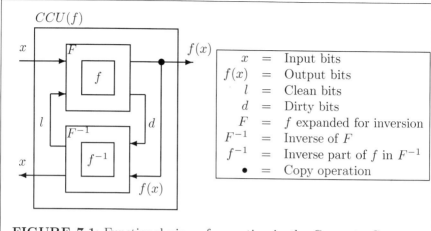

FIGURE 7.1: Functional view of execution in the Compute–Copy–Uncompute paradigm.

execution is found to be incorrect, and is re-executed. This type of "optimistic" execution finds use in systems such as parallel discrete event simulation [Carothers et al., 1999] and on-chip speculative execution [Dubois et al., 2012]. The execution of a program fragment is attempted optimistically even before all its pre-conditions about data dependencies are fully met. When any updates to its data dependencies are detected, the previous optimistic execution is reversed, and the program fragment is re-executed. This process is repeated, analogous to a fixed-point computation, until a guarantee is obtained that the dependencies will not change any further. At that time, the output effects of the code fragment are "committed" as final. Thus, the execution proceeds as a sequence of *forward* and *reverse* executions, ultimately ending in a forward execution that is finally *commit*ted.

When the execution is seen as a string of tokens, each token being a forward, reverse, or commit operation, the grammar satisfied by the execution string is shown in Table 7.2. Thus, for example, an execution trace in which the program fragment P is executed and reversed two times before finally being committed, the execution trace would be equal to $F(P) \rightsquigarrow R(P) \rightsquigarrow F(P) \rightsquigarrow R(P) \rightsquigarrow F(P) \rightsquigarrow C(F(P))$. All the possible combinations of the control flow are shown in the functional view of the execution in Figure 7.2.

7.3.1 Equivalence Conditions

Given that the program fragment P is executed using the Forward–Reverse–Commit paradigm, the results obtained from a traditional forward-only execution $\overline{F}$ equal the results from the reversible execution if the three new procedures, F, R and C, are such that $Output(\overline{F}(P)) = Output(FRC(P))$.

TABLE 7.2: Relaxation of Forward–Only Execution into the Forward–Reverse–Commit Paradigm

Forward–only	Forward–Reverse–Commit Execution
$\overline{F}(P)$	$FRC(P) \equiv [F(P) \rightsquigarrow R(P)]^* \rightsquigarrow F(P) \rightsquigarrow C(F(P))$

<u>Notation</u>
$$
\begin{aligned}
P &= \text{Program code fragment} \\
\overline{F}(P) &= \text{Traditional forward-only execution of } P \\
F(P) &= \text{Reversible forward execution of } P \\
R(P) &= \text{Reverse execution of } P \text{ after } F(P) \\
C(F(P)) &= \text{Committing to irreversibility of } F(P) \\
X \rightsquigarrow Y &= X \text{ followed by } Y \\
X^* &= \text{Zero or more executions of } X
\end{aligned}
$$

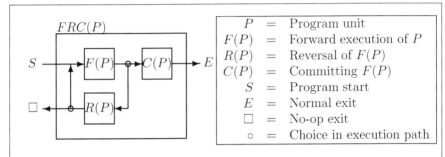

$$
\begin{aligned}
P &= \text{Program unit} \\
F(P) &= \text{Forward execution of } P \\
R(P) &= \text{Reversal of } F(P) \\
C(P) &= \text{Committing } F(P) \\
S &= \text{Program start} \\
E &= \text{Normal exit} \\
\square &= \text{No-op exit} \\
\circ &= \text{Choice in execution path}
\end{aligned}
$$

FIGURE 7.2: Functional view of execution in the Forward–Reverse–Commit paradigm.

Note that $\overline{F}(P)$ and $F(P)$ can be, in general, different from each other. They are equivalent only to the extent that the final outcome from the reversible execution satisfies the outcomes from the forward–only execution. In other words, $F(P)$ can be different from $\overline{F}(P)$, as long as the equivalence condition is satisfied by the chosen version of $F(P)$. Note also that, in general, the pair $\langle F(P), R(P) \rangle$ is not necessarily unique for a given $\overline{F}(P)$—multiple different pairs might exist that satisfy the equivalence condition.

The development of F, R, and C for a given P is illustrated in Section 13.3 with the problem of reversible dynamic memory allocation. A Forward-Reverse-Commit solution is developed and its correctness is proved by verifying the equivalence conditions.

7.3.2 Sequence Conditions

When $P = P_1 \rightsquigarrow P_2$ is a sequence of two code fragments, P_1 and P_2, then, this paradigm allows all interleavings of the operations contained within the individual expansions of $FRC(P_1)$ and $FRC(P_2)$, as long as the interleavings satisfy the following conditions:

1. The commit operation on P_1 must be performed before the commit operation on P_2. In other words, in the execution string, $C(P_1)$ must occur before $C(P_2)$. This ensures that the commit order duplicates the final sequence of effects of the forward-only execution of the $P_1 \rightsquigarrow P_2$ sequence.

2. The interleavings contain an odd number of occurrences of forward execution of P_1 before every forward execution of P_2. In other words, in the execution string, every occurrence of $F(P_2)$ must be preceded by an odd number of $F(P_1)$. This allows for any number of forward and reverse executions of P_1 to be performed, yet ensures that a forward execution of P_2 is not performed without having performed an un-reversed forward execution of P_1 prior to any forward execution of P_2. Note that it does not preclude any forward execution of P_2 from being later reversed.

7.3.3 Sequence Examples

Using the abbreviation $F_i = F(P_i)$, $R_i = R(P_i)$, $C_i = C(P_i)$, and $XY = X \rightsquigarrow Y$, the following are examples of valid and invalid sequences:

- $F_1 R_1 F_1 R_1 F_1 F_2 R_2 C_1 F_2 C_2$ is **valid**, equivalent to $F_1 C_1 F_2 C_2$.

- $F_1 R_1 F_1 R_1 F_1 F_2 C_1 R_2$ is **valid**, equivalent to $F_1 C_1$.

- $F_1 R_1 F_2 C_2$, is **invalid** because F_2 must be preceded by at least one F_1 that has not been reversed.

7.4 Undo–Redo–Do Paradigm

Consider interactive systems, such as graphical user interface-based applications, in which operations are performed in an error-prone, tentative, or exploratory manner, and need to be undone on demand. For example, a word processing program or graphical drawing application is expected to provide support for undoing certain actions initiated by the user. The reason for undoing could be because the action was from an unintentional error (e.g., typographical), or the action resulted in an unacceptable result (e.g., wrong color selection), or that action was taken to explore alternatives that will either be retained or rejected during use. In all such situations, the notion of reversible computing appears in the form of the ability to reverse the effects of

TABLE 7.3: Relaxation of Forward-Only Execution into the Undo–Redo–Do paradigm

Forward-Only	Undo-Redo-Do Execution
$\overline{F}(P)$	$URD(P) \equiv D(P) \rightsquigarrow [U(D(P)) \rightsquigarrow R(D(P))]^*$

Notation		
P	$=$	Program code fragment
$\overline{F}(P)$	$=$	Traditional forward-only execution of P
$D(P)$	$=$	Reversible forward execution (Do) of P
$U(D(P))$	$=$	Reverse execution (Undo) of P after $D(P)$
$R(D(P))$	$=$	Coast forwarding (Redo) of $D(P)$
$X \rightsquigarrow Y$	$=$	X followed by Y
X^*	$=$	Zero or more executions of X

program fragments. Reversible computing is realized in such cases using the Undo-Redo-Do paradigm.

In this paradigm, as a sequence of actions (*Do* operations) is performed, the actions are tracked and recorded so that, at any point in time, the sequence of actions can be retraced backward, undoing their effects in reverse order. Moreover, after some actions have been undone, *the act of undoing can itself be undone*, which is also called a *Redo* operation. Thus, a sequence of Undo operations can be followed by a sequence of Redo operations, which re-incorporate the original actions after they have been undone. Normal operation can resume by performing any new action (*Do* operation), which gets appended to the list of actions. If there are any undone actions that are not redone at the time a new action is added, the undone actions are purged and forgotten. This essentially discards the old chain of progress by pruning the Undo tree at the current branch point and starts a new branch with the new action.

When the overall execution is viewed as a string of tokens formed by $U(P)$ for Undo, $R(U(P))$ for Redo, and $D(P)$ for Do operations, the grammar of accepted execution is shown in Table 7.3. The flow of execution for the various stages of operation is shown in Figure 7.3.

Redo is also sometimes known as *coast-forwarding* because the system can advance the program state automatically to the desired position into the future, coasting along with little intervention or additional information needed from the user. The technique of coast-forwarding can sometimes be applied in the Forward-Reverse-Commit paradigm as well, when re-executing a previously reversed forward execution.

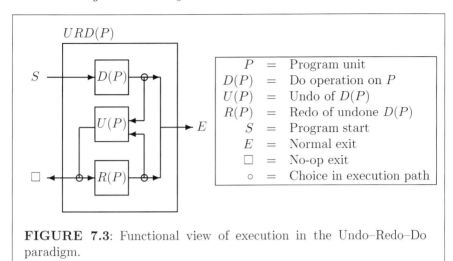

FIGURE 7.3: Functional view of execution in the Undo–Redo–Do paradigm.

7.5   Begin–Rollback–Commit Paradigm

This paradigm is often used in applications such as database transaction processing, and it is supported in some programming languages. Transaction processing in databases relies heavily on this concept to achieve atomicity and other important properties in executing a set of logically grouped operations. Programming languages such as FORTRESS [Allen et al., 2007, 2008] define similar rollback semantics using language constructs (e.g., abortable atomicity and nested transactions). The basic idea is for a flow of execution to be able to identify some important markers before it begins to make a group of operations that should be executed atomically, and ensure that either the entire group of operations completes successfully or the entire group becomes void (i.e., becomes as good as a no-op). The program itself can decide when any step has not succeeded, in which case it can initiate a rollback to the last marker or the runtime system also can initiate the rollback on the program's behalf if a runtime error is detected. The key is that the logical group of operations must be reversibly computed, so that their effects can be reversed as necessary at runtime.

Logically, every transaction starts the *Begin* stage, and rollbacks are initiated in the *Rollback* step, or all actions from the most recent *Begin* stage to the current point are committed in the *Commit* step. The functional view of this paradigm is shown in Figure 7.4.

The *Begin* stage signifies a marked point in the sequence of steps to which execution can be later rolled back. For example, in database transactions, this could either be the start of a new transaction, or it could be a marker

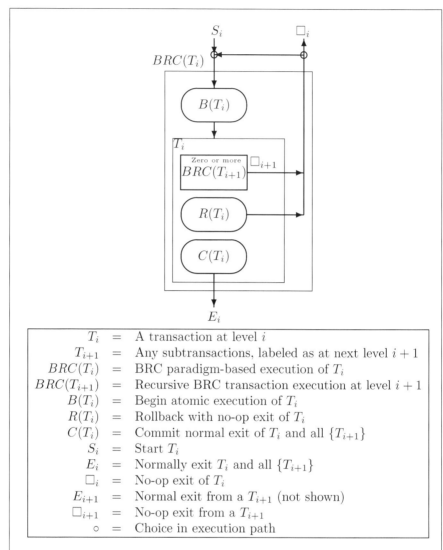

The following definitions accompany the figure:

T_i	$=$	A transaction at level i
T_{i+1}	$=$	Any subtransactions, labeled as at next level $i + 1$
$BRC(T_i)$	$=$	BRC paradigm-based execution of T_i
$BRC(T_{i+1})$	$=$	Recursive BRC transaction execution at level $i + 1$
$B(T_i)$	$=$	Begin atomic execution of T_i
$R(T_i)$	$=$	Rollback with no-op exit of T_i
$C(T_i)$	$=$	Commit normal exit of T_i and all $\{T_{i+1}\}$
S_i	$=$	Start T_i
E_i	$=$	Normally exit T_i and all $\{T_{i+1}\}$
$\square_i$	$=$	No-op exit of T_i
E_{i+1}	$=$	Normal exit from a T_{i+1} (not shown)
$\square_{i+1}$	$=$	No-op exit from a T_{i+1}
$\circ$	$=$	Choice in execution path

FIGURE 7.4: Functional view of execution in the Begin–Rollback–Commit paradigm.

within the transaction. In general, it is theoretically possible to rollback to any specific point in the past. However, typically, syntactic constructs (e.g., a `BEGIN TRANSACTION` or `SAVEPOINT` statement in the Structured Query Language (SQL) for databases) is used to make the marked point explicit. A *Rollback* stage can later be initiated at any time, to abandon all side effects of the steps from the previous *Begin* stage until the current point of execution, erasing any and all modifications initiated as part of the current transaction. Again, syntactic constructs or other interfaces can be used to initiate the rollback (e.g., a `ROLLBACK TRANSACTION` statement in SQL). The *Commit* operation serves as the only indicator that the steps in the transactions because the previous *Begin* stage will not be subject to rollback, and hence any and all side effects, such as reads and writes to the data, must be pushed to the actual locations where data is stored. Any later transactions will be able to access the modified data in a consistent and correct manner.

The power of this execution model is enhanced even further by allowing for nested transactions [Resende and Abbadi, 1994]: any of the steps in a transaction may begin another *subtransaction* that also supports all the same semantics as the parent transaction, and propagate errors, if any, to its parent. If a parent transaction T_i at level i starts a subtransaction T_{i+1} at level $i+1$, any reversal initiated on T_{i+1} gets propagated up the chain to T_i, which is also consequently rolled back. The transaction chain starts at T_1 and finishes with a normal exit E_i if T_i and the set of all its subtransactions $\{T_{i+1}\}$ terminate normally. T_i ends with a no-op exit $\square_i$ if itself or any of its subtransactions initiates a rollback.

Part III

Software

Chapter 8

Reversible Programming Languages

8.1 Role of Reversible Languages

Traditional programming languages contain a mix of reversible and irreversible constructs. Thus, in general, programs written in traditional languages can be reversible or irreversible, depending on the constructs they use. If a program uses even a single irreversible construct, the program as a whole can become irreversible. Irreversible programs can indeed be reversibly executed using simulation techniques to map irreversible execution to reversible platforms, but at the cost of added runtime and memory overheads. In other words, the runtime and memory costs of reversible execution are (polynomially) larger than those of irreversible execution. It is conceivable to develop a preprocessor to detect when a program is free of irreversible statements and compile such a program differently to avoid the simulation overheads in enabling reversibility. On the other hand, new *reversible* programming languages that provide only reversible constructs avoid all runtime overheads for reversibility.

IP =   Irreversible programs written using irreversible languages can be reversibly executed via simulation of irreversible programs over reversible platforms, but with runtime overhead incurred for the simulation.

IRP =   Reversible programs written using a subset of irreversible languages, reversibly executable by special detection and compilation for reversible platforms; the same programs can also be expressed using reversible languages and can be reversibly executed directly; both approaches can avoid incurring runtime overheads.

RP =   Reversible programs that can be expressed only with reversible languages to avoid all runtime overheads for reversible execution.

FIGURE 8.1: Classes of irreversible and reversible language programs.

With these considerations, we have three classes of programs, as shown in Figure 8.1: (1) IP are programs written in irreversible languages using one more irreversible constructs, and hence necessarily incur simulation costs to enable reversibility; (2) IRP are programs written either in irreversible languages but using only the restricted subset of reversible constructs, or programs that are written using reversible languages; and (3) RP are programs written in reversible languages that cannot be expressed with irreversible languages. Examples in the RP class are those using reversible floating point arithmetic operations or reversible memory operations such as swaps that are not available in conventional languages. Reversible programming languages are hence useful to not only provide reversible execution with assured efficiency (e.g., memory usage not increasing with execution length), but also to allow the programmer to more naturally express algorithms that are reversible by design.

Most conventional programming languages such as **Fortran**, **C**, and **C++** can be used to write irreversible programs that can be augmented with run-time instrumentation via compiler techniques to enable simulation of their execution for reversibility. It also is possible to identify a reversible subset of each irreversible language to define a sub-language that is reversible, although the expressive power of the subset may significantly limit the types of programs that can be written.

Several reversible languages have also been defined over the past few decades, with varying levels of expressive power, variety of constructs, and ease of programming. These include **Janus**, **R**, **EPA**, **SRL**, **ESRL**, **PSIL-ISP**, **INTERLISP**, **Inv**, and **pGCL**.

8.2 Language-Level Reversibility Issues

For a programming language to provide reversibility of execution, conventional notions in control flow and primitive operations must be redefined due to their irreversible nature of operation. The irreversible notions arise in the form of destructive updates to variables, conditional branching statements that lose information, loops with variable number of iterations, and lossy arithmetic operations. In reversible programming languages, new constructs are defined to eliminate reversibility problems that arise due to each of these issues, which are fundamental in nature although they might manifest themselves in different forms due to syntactical differences among different languages. The overall effectiveness of the reversal directly depends on how each of these issues is treated to minimize the overheads introduced for reversibility.

8.2.1 Sequence and Recursive Reversal

A program is a sequence of statements $< S_1, \ldots, S_n >$ expressing the forward control flow of execution. The reversal of the sequence is the inverted forward sequence containing the inverses of each individual statement: $< S_n^{-1}, \ldots, S_1^{-1} >$. Each statement S_i may be any of the other types of statements, such as destructive updates, conditional statements, looping statements, and so on, whose reversal is described in the next subsections.

8.2.2 Destructive Updates

A destructive update of a variable is any statement of the form

$$x \leftarrow E(x', \mathcal{V}),$$

where x is the variable (named memory location) being overwritten, x' is the pre-assignment value of x, e is an expression over x' and other variables $\mathcal{V}$, and the expression E is such that no E^{-1} can be found that recovers the pre-assignment value of x as $x' = E^{-1}(x, \mathcal{V})$. The simplest example of a destructive update is an assignment of the form $x \leftarrow y$, where y is another variable whose current value is not functionally related to the pre-assignment value of x. A further generalization of destructive updates can be obtained by expanding the assignment to a set of variables: $X \leftarrow \mathcal{E}(X', \mathcal{V})$, where X is a composite of one or more variables being updated by a group of executable statements represented by $\mathcal{E}$, X' is the pre-assignment value of X, and $\mathcal{E}^{-1}$ is a set of executable statements that can reverse the effects of $\mathcal{E}$ to restore X to X'. All the assignments to X for which no $\mathcal{E}^{-1}$ can be found constitute an aggregate destructive update.

In contrast, *constructive updates* provide an invertible, functional relation between the pre-assignment value and the variables used to compute the new value. The simplest example is the swap operation of the form $x \leftrightarrow y$, which can be easily inverted by the same operation $y \leftrightarrow x$ that swaps back the values of x and y. Destructive updates are opposite of constructive updates.

Reversible languages must prohibit any expression e that is not possible to symbolically invert, and restrict updates to invertible operators. For example, only invertible operations such as swap, increment, and decrement may be defined by the language, and the language syntax may be defined to make it impossible to express any destructive update. Any additional programming complexity may be obtained from compositions of such specifically designed invertible arithmetic operators.

8.2.3 Arithmetic

In conventional computers, arithmetic is in general irreversible. To ensure reversibility, arithmetic-based modifications must be restricted to be *constructive* in nature, such as simple accumulations or decrements (`+=` and `-=`). More general solutions include relaxations and redefinitions of multiplication and division operations to eliminate all possibilities of overflow and underflow that lose information. For example, the semantics of multiplication may be redefined to employ three operands instead of two operands, to address all sources of irreversibility. Modulo arithmetic is another important component of reversible arithmetic.

8.2.4   Conditionals

Branches of control flow due to conditionals have the potential to introduce irreversibility because they can create ambiguity on the backward path at the point at which forward branches converge. Conditional statements introduce a split of control flow into two or more branches, which are then merged into a single flow later in the execution. When executing in reverse, information is needed at the merge point as to which of the branches must be traversed in reverse. The ambiguity cannot be resolved simply by reevaluating the forward condition. This is because, if any of the variables appearing in the condition expression is modified inside the branches of the conditional statement, then the condition expression is not guaranteed to give the same value as its value from forward execution, and hence becomes irreversible. Even in the simplest case of reversing a conditional statement with only two branches (true and false branches), in general, it is impossible to recollect which of the bodies, the true branch or the false branch, was executed in the forward path.

For example, a conditional statement such as

$$\textbf{if } x < 0 \textbf{ then}$$
$$x \leftarrow -x$$
$$\textbf{else}$$
$$x \leftarrow 2 * x$$
$$\textbf{end if}$$

cannot restore the correct pre-modification value of x because x is being modified inside the body of the **if** statement. If x is a positive even number after the branch is executed, then the statement cannot be uniquely reversed, because all negative even numbers and all positive (even or odd) numbers are mapped to positive numbers by the branch statement. There is sufficient information to reverse correctly if x happens to be an odd number (which would guarantee that the true branch was taken in the forward mode, as the false branch always converts the number to become even). However, the ambiguity remains for even numbers, and hence the statement in general cannot be reversed correctly.

In general, consider the conditional statements in the following form:

$$\textbf{if } e \textbf{ then}$$
$$S_1$$
$$\textbf{else}$$
$$S_2$$
$$\textbf{end if}$$

Let the set of variables being used in the condition expression e be $R(e)$. Let the set of variables being overwritten in the statements S_1 in the true branch be $W(S_1)$ and, similarly, that of the false branch be $W(S_2)$. Then the reversibility of the conditional statement can be violated if $C1 = R(e) \cap W(S_1) \neq \varnothing$ or $C2 = R(e) \cap W(S_2) \neq \varnothing$.

Indeed, if $C1 = \varnothing$ and $C2 = \varnothing$, then the branching decision can be recovered correctly by simply reevaluating the same forward expression e, and the branch statement can be correctly inverted as follows:

$$\textbf{if } e \textbf{ then}$$
$$S_1^{-1}$$
$$\textbf{else}$$
$$S_2^{-1}$$
$$\textbf{end if}$$

The reverse statements S_1^{-1} and S_2^{-1} are determined by recursively applying the reversal approach on the corresponding statements S_1 and S_2, as indicated in Section 8.2.1.

In reversible programming languages, there are essentially three ways to address the problem of reversibility of conditional statements:

- ■ **Restriction of Semantics** Because the irreversibility disappears if $C1 = \varnothing$ and $C2 = \varnothing$, one way to ensure reversibility is to prohibit the program from having a non-empty set for either $C1$ or $C2$. It is left to the programmer to rewrite the program in such a way that this condition is ensured at runtime; violation of the reversibility requirement would lead to undefined results.

- ■ **Post-condition** Post-conditions are a common technique by which the difficulty of reversing conditionals is addressed. In this approach, in addition to the usual branching condition, which we will call the pre-branch expression or e_{pre}, a new condition, called the post-branch expression or e_{post}, is introduced to the syntax of the branch statement. The semantics require that, in forward execution, the e_{post} expression at the end of the branch statement must evaluate to the same truth value to which e_{pre} evaluated before the branch was taken. This ensures that the reverse mode can rely on e_{post} to determine which branch should be reversed.

Forward	Reverse
$\textbf{if } e_{pre} \textbf{ then}$	$\textbf{if } e_{post} \textbf{ then}$
S_1	S_1^{-1}
$\textbf{else}$	$\textbf{else}$
S_2	S_2^{-1}
$\textbf{end } e_{post} \textbf{ if}$	$\textbf{end } e_{pre} \textbf{ if}$

Languages such as **Janus** require a post-condition for a simple `if` statement. Generalization to multi-way conditional statements, such as the `switch` statement, was defined in [Glück and Kawabe, 2003], where a unique post-condition e_{post_i} is required to be defined for every branch i, with no mutual intersection after the end of the multi-way statement (i.e., $e_{post_i} \cap e_{post_j} = \varnothing, \forall i, j$).

■ **Pragma Specifications** This approach is a trade-off between the previous two solutions. The programmer is required to specify to the compiler one of three options: (1) there is no reversibility hazard, that is, $C1$ and $C2$ are both assured to be null sets; or (2) a post-branch expression is specified that can recover the pre-branch expression's truth value; or (3) the compiler must generate a temporary bit variable that records the forward pre-branch truth value, which should be used in the reverse execution.

The default could be the last option, so if the programmer does not specify the first or second, the compiler automatically inserts the bit needed to remember the branching decision. The language can permit the specification of the programmer's option via a pragma interface. For example, `#pragma POST_CONDITION "INVARIANT"` just before the `if` statement applies the first option to that statement, while `#pragma POST_CONDITION` *expression* specifies the second option.

8.2.5   Loops

Looping with a variable number of iterations is a fundamental and powerful programming construct through which much of the complexity arises in most software (in fact, it is so powerful that the halting problem would disappear if variable iteration is prohibited in procedural languages). In general, the power comes at the cost of reversibility in conventional programming languages. The crux of the problem is that, without taking careful measures, the exact count of the number of iterations taken in the forward mode cannot be recovered when executed in reverse. Here, we illustrate reversibility issues in the context of iterative loops of the following form (other types of loops can be treated in an analogous manner):

$$\textbf{while } e \textbf{ do}$$
$$S$$

In reversible programming languages, there are essentially two ways to address the problem of reversibility of looping statements:

1. **Post-condition** In this approach, instead of a single looping condition, the looping template is relaxed to include two conditions: the pre-loop expression or e_{pre}, and a post-loop expression or e_{post}. The semantics require that, in forward execution, the e_{pre} pre-loop expression must be true before the loop but false after the loop starts. Similarly, the e_{post} post-loop expression is false until the loop ends. This ensures that the reverse mode can rely on e_{pre} to determine when the loop must be

terminated in reverse.

Forward	Reverse
assert e_{pre}	assert e_{post}
while not e_{post} do	while not e_{pre} do
S	S^{-1}
assert not e_{pre}	assert not e_{post}

2. **Counter-Based Reversal** In this approach, a new temporary variable is introduced by the compiler to remember the number of times the loop body was executed in the forward mode.

Forward	Reverse
$c \leftarrow 0$	
while e do	while $c > 0$ do
S	S^{-1}
$c \leftarrow c + 1$	$c \leftarrow c - 1$

3. **Pragma Specifications** A language interface can be used by the programmer to specify a symbolic stopping condition expression that the reverse mode execution can use to terminate the loop during execution. The compiler inserts this expression into the reverse loop. Alternatively, the name of a variable that maintains the loop count in the program can be specified, so that no new variable needs to be introduced by the compiler. If the programmer does not specify either option, the compiler automatically inserts the counter variable needed to remember the loop count. The language can enable the specification of the programmer's options via a pragma interface. For example, `#pragma LOOP_CONDITION` *expression* just before the `while` statement applies the first option to that statement, while `#pragma LOOP_COUNTER` *variable* specifies the second option, and `#pragma LOOP_COUNTER "AUTOMATIC"` specifies the third option.

4. **Restriction of Semantics** In this approach (which may be considered draconian), no support for variable iteration is provided. Loops with a fixed number of iterations are allowed. Although highly restrictive in nature, this approach does still allow many algorithms to be effectively expressed in terms of loops with fixed iteration counts. Algorithms such as sorting and Fourier transforms can indeed be expressed under such restriction of semantics, with the added assurance of reversibility. Some reversible languages have been defined in the literature with such a constraint and used for the purposes of theoretical analyses.

8.2.6 Additional Control Flow Issues

Additional issues need to be similarly addressed, and reversibility of advanced control flow constructs remains an open research as of this writing.

■ Jump instructions such as `goto` statements can be made reversible by different solutions. One way is the introduction of a reverse-peer instruction such as `comefrom`. Another method is to prevent statements from being targets of more than one jump instruction. Yet another method is to have the compiler generate certain disambiguating variables that keep track of the origin of the jump when the jump is potentially made from multiple origin points.

■ Reversal of procedure calls is achieved by one of two ways, depending on the specific language support used. In languages in which every statement is reversible, there is only one instance of the procedure that can be executed in forward or reverse mode. Moreover, in some languages, a procedure can in fact be called or "uncalled" in forward mode itself. The corresponding calls become reversed so that the procedure becomes uncalled or called, respectively, in reverse mode. In languages that generate two separate instances of the procedure, one for forward execution and the other for reverse execution, the invocation of a (forward instance) procedure call in forward mode gets reversed by invocation to the reverse instance of the procedure.

■ Recursion is naturally handled by the same techniques as for a subroutine call.

■ Exception handling, such as the `try-catch` statement in C^{++}, is possible to address via normalization to basic control flow such as branching and jump instructions, but a more elegant, high-level reversal remains to be found.

■ Multi-threaded execution similarly remains to be addressed, the thorny issue being the compromise among multiple conflicting issues such as efficiency, determinism, and fairness, all of which come into play when reversibility is considered.

8.3 Procedural Languages

Procedural languages make the largest class of programming languages by a very wide margin. However, almost all existing procedural languages have been designed for forward-only execution. Conventional forward-only procedural programming enjoyed many decades of research and development in forward execution languages. In comparison, there has been relatively little development with respect to reversible execution languages. Perhaps largely due to the lack of sufficient motivation for reversibility, programming language designers have seldom incorporated any consideration for reversibility. Many other considerations, such as runtime performance, code reuse, reduction of unintentional errors, and the like, have dominated programming language development. However, the concept of reversibility of execution introduces an entirely new dimension in language design, orthogonal to almost all traditional considerations in programming languages.

Imagine the set of issues and challenges that arise in defining a programming language with reversible execution as the primary consideration. To begin with, the simplest yet thorny issue is the definition of data types of variables and reversible operations on those data types. The data types and operations must provide sufficient expressive power, yet avoid loss of information in all cases, *by design*. Power of expression and programmability need to be maintained while accommodating reversibility. Similarly, all control flow constructs must be reversible by design, yet support complex logic. Because modularity features such as subroutines are needed for software engineering purposes, they too need be enhanced with appropriate concepts of reversibility.

Here, a few recently proposed reversible procedural languages are described that are intended to provide *reversibility by design*. Additionally, the issue of *retroactively* introducing reversibility concepts into existing (forward-only) procedural languages is also considered in the context of the popular **C** language.

8.3.1   SRL and ESRL Reversible Languages

Let us first focus on languages with some of the simplest instruction types and consider their reversibility properties. The set of functions computable by Turing Machines is the same as that defined by partial recursive functions. A subset of partial recursive functions is the set of primitive recursive functions, which are defined on natural numbers with a bounded number of computations/iterations. Consider a language to express primitive recursive functions, defined on non-negative integers, in which the only instructions allowed are of three types: increment (INC *var*), decrement (DEC *var*), and fixed iteration (FOR *var* *P*, where *P* is any subprogram written in the language). The operations are defined on a fixed set of variables (e.g., multiplication of two numbers a and b can be achieved via an increment operation inside two nested FOR loops on a and b). This language is, for the most part, easy to reverse, because increment and decrement are mutual inverses, and fixed iteration is reversed by invoking the inverse of its body. However, because the variables are only allowed to take on non-negative values (the decrement operator is defined to enforce this), information loss arises whenever a decrement operation occurs on any variable that is currently at zero. Thus, even the set of primitive recursive functions is irreversible.

The limitation of primitive recursive functions with respect to reversibility can be solved by relaxing the variables to assume arbitrary integer values (not just non-negative values). In other words, instead of the set of natural numbers $\mathcal{N}$ as the domain, the set of integers $\mathcal{Z}$ can be used as the domain. Reversibility is then restored in this new set of functions, defined in [Matos, 2003] as the Simple Reversible Language (**SRL**) language.

However the new set of functions defined by **SRL** on integers has the limitation that certain operations cannot be achieved without sideeffects. For example, changing the sign of the value in a variable cannot be achieved without lasting modifications to at least one other variable. Similarly, values in two variables cannot be swapped without modifications to at least one other variable. However, these limitations are overcome by the addition of one additional instruction type, namely, the negation operator: `NEG` *var* changes the sign of the variable *var*. This extension of **SRL** is defined in [Matos, 2003] as the Extended Simple Reversible Language (**ESRL**). Languages such as **SRL** and **ESRL** are useful for theoretical purposes, to help understand the basic abilities and inabilities of even the simplest procedural languages when considered in the context of reversibility.

8.3.2 EPA Reversible Language

The Event-Predicate-Action (**EPA**) language is an early reversible language designed to enable automatic and efficient reversible execution of event-oriented software [Briggs, 1987]. Within a runtime infrastructure that accepts events from the user environment, evaluates a set of predicates to determine which predicates become satisfied by the newly arriving events, and fires off actions that are activated by the predicates that guard those actions. The actions are entirely composed of reversible statements to enable any action to be undone for different reasons such as invalidation of an event, or an adjustment of the data associated with an event, or a correction of the order of events. Thus, the crux of reversibility lies in the ability to reverse the statements of the software language used to program the actions. A code generator accepts **EPA** programs and emits forward and reverse versions of every action, both operating on a shared set of variables that maintain a consistent state of the application.

EPA defines two types of assignments to variables, one constructive in nature and the other destructive in nature. The difference between the two is to permit reduction of the memory needed for reversal: constructive updates can be reversed by computing their corresponding inverses, while only destructive updates need memory to save the pre-assignment values being overwritten. As constructive updates, only the increment operation is permitted on integer data types. Destructive updates include initialization as well as overwriting. Destructive updates save pre-assignment values. Care is taken to avoid saving and restoring initialization values because the reversal of initial assignments is immaterial to the correctness of action reversal. Conditional statements are reversed by saving a single bit to record the branching decision. Loops are reversed using a post-condition approach. Overall, a large reduction in the amount of memory was achieved for reversal of a real-life program, namely an application for cricket scoreboard maintenance. The reduction was achieved to the large fraction of constructive updates and due to careful minimization of the number of occasions at which values are saved to memory. Gains in software development, debugging, and testing efforts were observed. The use of a reversible language and the employment of a compiler to provide automated reversible computation helped achieve a reduction in development time, assurance of correctness, minimization of memory usage, portability to different computing platforms, and increased maintainability of the software.

8.3.3   Janus Reversible Language

Among the earliest procedural languages for general-purpose reversible programming, the **Janus** language [Lutz and Derby, 1986] is perhaps the most comprehensive and elegant creation. It is a very compact and beautifully defined language, with a very simple syntax. The semantics are restricted by design to ensure reversibility at all levels. The language was cleverly designed to avoid irreversible features, yet provide sufficient power to develop complex algorithms. While the authors of the language originally intended the language to be a preliminary exercise in reversible language design, the language was later shown to be expressive enough to describe a Reversible Turing Machine using the language constructs alone, in only a few dozen lines of **Janus** code [Yokoyama, 2010]). The entire grammar of the language is shown in descriptive form in Table 8.1. A brief form of the **Janus** syntax is provided in the original article [Lutz and Derby, 1986] that first introduced the language. An even more succinct form of the same was published in [Yokoyama and Glück, 2007].

   A **Janus** program consists of a list of variable name declarations and a list of procedure (subroutine) definitions that use the variables in the procedure bodies. The procedure bodies contain executable statements, each of which is reversible by design. Executable statements can be sequences of statements, swap statements, (constructive) assignment statements, conditionals, loops, procedure calls, and input/output operations. Control flow is carefully defined to always preserve reversibility. The most basic control flow statement is the swap (or exchange) binary operator (:) on two variables, which simply swaps the values in the operands. Clearly, it is a self-inverse operator. Except for the swap operator, all the remaining control flow constructs have unique characteristics, making these constructs qualitatively quite distinct from those in conventional forward-only languages. These are described in detail next.

8.3.3.1   Data Structures and Reversible Arithmetic

   The data types are limited to integers, and the data structures are limited to scalars and arrays. Array subscripts can be integer expressions. Integers are permitted to assume only non-negative values up to a specific number of bits (e.g., 32-bit), and all arithmetic is performed modulo the maximum value. All

TABLE 8.1: Grammar of the **Janus** Time-Reversible Language

Element		Syntax	Notes
program	:=	*declare procedures*	Variable declarations and procedures
declare	:=	*name* | *name declare*	List of variable names
procedures	:=	*procedure procedures* | *procedure*	One or more procedure definitions
procedure	:=	PROCEDURE *name stmt*	A procedure definition
stmt	:=	*stmt stmt* | *swap* | *assign* | *if* | *loop* | *call* | *input* | *output* | SKIP	Sequence, swapping, assignment, conditional, loop, procedure call/uncall, input, output, or no-effect statement
if	:=	IF *condition*^{pre} THEN *stmt*^{if} ELSE *stmt*^{else} ENDIF *condition*^{post}	Truth value of *condition*^{pre} evaluated before execution of *stmt*^{if} or *stmt*^{else} should be equal to truth value of *condition*^{post} evaluated after execution
loop	:=	FROM *condition*^{from} DO *stmt*^{do} LOOP *stmt*^{loop} UNTIL *condition*^{until}	Assert that *condition*^{from} is true, execute *stmt*^{do}, and then, repeatedly execute *stmt*^{loop} until *condition*^{until} becomes true
call	:=	CALL *name* | UNCALL *name*	Call or uncall procedure *name*
swap	:=	*var*^{1} : *var*^{2}	Swap values of *var*^{1} and *var*^{2}
assign	:=	*var* += *expression* | *var* -= *expression* | *var* ^= *expression*	Evaluate *expression* and add to, subtract from, or bit-wise exclusive-or with *var*
expression	:=	⊙ *expression* | *expression* ⊗ *expression* | *var* | integer	Unary (⊙) and binary (⊗) operations, variables, and constants
var	:=	*name*[*expression*] | *name*	Array or scalar variable
input	:=	READ *name*	Swap the value of *name* with a user-given value from the environment
output	:=	WRITE *name*	Swap the value of *name* with a user-given location in the environment

variables are guaranteed to be initialized to zero before program execution Non-zero values can be inserted initially into the program via input operation to assign values from the system to specific variables of the program.

Only reversible arithmetic is defined on integers, as *constructive assignment* operations of the form *lvalue operator rvalue*. The only operators supported are *add to* (+=), *subtract from* (-=), and *exclusive bit-wise or with* (^=) any integer expression. In a constructive assignment, the integer variable being modified is the left-hand-side reference (an *lvalue*), and the integer expression being used as an operand is the right-hand-side value (an *rvalue*). It is constructive in the sense that modifications never lose information, making it possible to *locally* reverse the assignment by reevaluating the *rvalue* and use it in restoration of the *lvalue* to its old value. Clearly, += and -= are mutual inverses, while ^= is a self-inverse.

8.3.3.2   Reversible Conditional Statement

Conditional (or branching) statements in conventional forward-only language constructs possess an asymmetry between pre-condition and post-condition actions. There is a natural notion of a condition or guard defining the state prior to entry of the statement, but there is no corresponding notion for the exit of the statement. **Janus** resolves this asymmetry for conditional statements by adding a new concept of post-condition and appropriate semantics that retain conventional forward-only expressive power while making it naturally symmetric and reversible.

The syntax and semantics of a **Janus** conditional statement are shown in Algorithm 8.1. In the algorithm, e_1 and e_2 are logical expressions. The distinction from a conventional forward-only construct is evident in the form of the expression e_2 at the end of the statement. The reversible execution semantics specify that the truth value of the pre-condition e_1 evaluated (in forward execution) before the conditional branch is taken must be equal to the truth value of the post-condition e_2 evaluated after the branch is executed. The semantics of this behavior are part of the language specification; the semantics are either assumed to be obeyed by the program or are enforced by the language runtime system. In other words, either execution proceeds with the assumption that the program somehow ensures that e_1 and e_2 correctly evaluate to the same truth value pre- and post-execution, respectively, or the runtime system raises a runtime exception if/when the requirement is violated by the program.

To help readers who are conversant with the traditional **C** language, the semantics of the reversible conditional statement are expressed in an equivalent **C** code fragment in Algorithm 8.1. In the **C** code, the assert(condition) call raises a runtime error if condition evaluates to false.

Algorithm 8.1 Reversible execution semantics of conditionals in **Janus**

	Forward	Reverse
Janus	IF e_1 THEN S_1 ELSE S_2 FI e_2	IF e_2 THEN S_1^{-1} ELSE S_2^{-1} FI e_1
	$\Downarrow$	$\Downarrow$

C	`int v = `e_1`;`
	`if(v) `S_1
	`else `S_2
	`assert(v == `e_2`);`
		`int v = `e_2`;`
		`if(v) `S_1^{-1}
		`else `S_2^{-1}
		`assert(v == `e_1`);`

8.3.3.3 Reversible Looping Statement

Similar to conditional statements, conventional looping statements (with variable number of iterations) are asymmetric in the sense that only an exit condition is explicitly defined, but there is no explicitly defined entry condition. **Janus** resolves this asymmetry for looping statements by adding a concept of entry condition and appropriate semantics that retain conventional forward-only expressive power while establishing a new symmetry for reversibility.

The definition of a reversible version of an iterative loop is nontrivial. The challenge is that the looping statement may execute for a variable number of iterations, determined at runtime by a user-defined loop exit condition. One of the most elegant parts of **Janus** is its definition of a new looping construct that enforces reversibility by design. The syntax and semantics of a **Janus** looping statement are shown in Algorithm 8.2. The operation of this looping construct is unconventional and complex. To help understand its operation, the equivalent expressions of its forward and reverse behavior are shown in **C** language in Algorithm 8.2. Also listed is an example to illustrate the sequence of evaluations of conditions and statements in forward and reverse execution of the loop. Note the symmetry of the conditions e_1 and e_2 with respect to forward and reverse modes.

A key enabler of reversibility for the **Janus** loop construct is the relation between the conditions e_1 and e_2. In the forward mode, e_1 must be true for entry into the loop but it must be false at the end of every iteration. The behavior is defined this way to ensure that the reverse mode can rely on e_1

as the stopping condition of the reversed loop, just as the forward mode relies on e_2 as the stopping condition of the forward loop.

It might appear as though this type of construct may be difficult to use in a program, but the authors of **Janus** have demonstrated its natural use in a few nontrivial problems such as the square root computation and sorting procedures.

Algorithm 8.2 Reversible execution semantics of looping in **Janus**

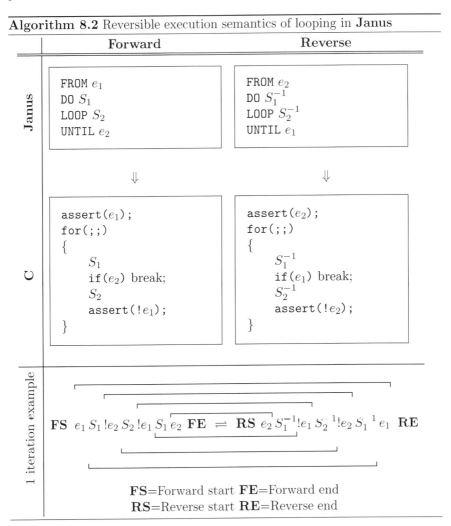

	Forward	Reverse
Janus	FROM e_1 DO S_1 LOOP S_2 UNTIL e_2	FROM e_2 DO S_1^{-1} LOOP S_2^{-1} UNTIL e_1
	$\Downarrow$	$\Downarrow$

C	`assert($e_1$);`
	 `for(;;)`
	 `{`
	 $\quad S_1$
	 `    if($e_2$) break;`
	 $\quad S_2$
	 `    assert(!$e_1$);`
	 `}`
		`assert($e_2$);`
		 `for(;;)`
		 `{`
		 $\quad S_1^{-1}$
		 `    if($e_1$) break;`
		 $\quad S_2^{-1}$
		 `    assert(!$e_2$);`
		 `}`

1 iteration example

FS $e_1\,S_1\,!e_2\,S_2\,!e_1\,S_1\,e_2$ **FE** $\rightleftharpoons$ **RS** $e_2\,S_1^{-1}\,!e_1\,S_2^{-1}\,!e_2\,S_1^{-1}\,e_1$ **RE**

FS=Forward start **FE**=Forward end
RS=Reverse start **RE**=Reverse end

8.3.3.4 Reversible Subroutine Invocation

Subroutines in all conventional languages are defined with a fundamental assumption of forward-only execution. When a subroutine is invoked, the caller always implicitly assumes that the control flow is transferred to the top of the subroutine and flows down to the end of the subroutine. Due to unidirectionality of execution, there is no scope for any other notion. To this conventional notion of *calling* a subroutine, **Janus** introduces the complementary concept of *uncalling* the same subroutine. In fact, this concept of calling or uncalling a callee subroutine is orthogonal to the direction of execution of the caller itself. In other words, the caller may be in forward or reverse mode of execution when calling or uncalling a subroutine. This is further elaborated shortly.

Subroutines are supported in **Janus** as procedures that can be called from other procedures. Procedures are allowed to be directly or indirectly recursive. Forward references to procedures are allowed before they are defined. The crucial difference between conventional subroutines and **Janus** subroutines is in the introduction of the additional concept of *uncall*ing, which is the reverse counterpart to the conventional forward invocation. In all conventional programming when a subroutine is called, control begins at the top of the subroutine and flows down to its end. In **Janus**, this conventional calling is available as usual, using the primitive CALL *procedurename*. In addition, the same procedure can also be uncalled using the primitive UNCALL *procedurename*. When the procedure is uncalled, control begins at the end of the subroutine and flows backward toward the beginning; moreover, in this uncall mode, the *inverses* of the statements are executed.

Consider a procedure P1 (caller) invoking another procedure P2 (callee). Depending on whether the caller is executing in forward or reverse mode and whether the callee is being called or uncalled, four different cases arise, as listed in Table 8.2. Note that this implies that a procedure can be uncalled from a procedure even in normal (forward) mode, and can be called in reverse mode. For example, if CALL *name* is used as forward execution, then UNCALL *name* becomes the reverse, but it is also possible to use UNCALL *name* in forward execution, in which case CALL *name* becomes the reverse.

TABLE 8.2: Reversible Calling Semantics of Subroutines in **Janus**

	Callee Mode	
Caller mode	*Forward*	*Reverse*
Forward	CALL	UNCALL
Forward	UNCALL	CALL
Reverse	CALL	UNCALL
Reverse	UNCALL	CALL

Although one may wonder why one would consider uncalling a procedure in a conventional program, there are indeed a few natural uses. Because there is no such concept in traditional programming, there is insufficient awareness of the possibilities; however, with more software development using reversible languages, the concept of uncalling can become more natural over time. The authors of **Janus** demonstrated the usage of uncalling in a square root computation and in a factorization program [Lutz and Derby, 1986].

8.3.3.5   Reversible Input and Output

Introduction of values into the program from the user/system (e.g., for initialization) is supported via a reversible input statement. Similarly, presentation of computed values from the program to the user/system (e.g., for printing answers) is supported via a reversible output statement.

In conventional forward-only execution, input and output operations are irreversible in nature. For example, a user may type some numbers at the computer terminal as input to the program, which is received into a memory variable via an input statement in the program. Reversal of such an input action is irreversible in two ways: (1) in forward mode, the value of the variable into which the input value is received would lose its old value, thereby potentially destroying some information; (2) in reverse mode, it is not clear how to define the return of the input value back to the user. Analogous issues arise with output from the program.

The reversibility issues from conventional input and output semantics are solved by **Janus** through clever alternative specifications of behavior that make the operations reversible by definition. In the new definition of input, a READ *var* statement means that the current value of the variable *var* in the program is exchanged with an unspecified variable in the program environment. If this READ statement is executed in reverse, the system is expected to supply the saved value back to the program while the program promises to return the input value back to the environment. Essentially, the input operation becomes an exchange or swap operation on an unspecified location of the system, delegating to the system the responsibility of restoration of the old value of the variable. An output statement WRITE *var* has analogous semantics: the program exchanges the value of the program variable *var* with an unspecified variable in the system. It can be seen that the reverse semantics of input are the same as the forward semantics of output, and vice versa. Thus, the reverse of READ *var* is WRITE *var*, and vice versa.

8.3.3.6   Global Reversal

In **Janus**, globally reversible execution is obtained by composition of local reversibility. Because every individual statement is guaranteed to always have an inverse, and that inverse is locally defined (solely dependent on that statement alone), reversibility at any level (e.g., procedure-level) is simply obtained by reversing below that level. The rules of reversal are shown in Table 8.3, which, when applied recursively, can be used to obtain a reversible execution for any statement.

Table 8.3 shows a summary of the instructions and their inverses. Note that there is no specific distinction as forward and reverse for the instruction

TABLE 8.3: Reversal of the **Janus** Language Instructions

Instruction	Inverse
$stmt_1\ stmt_2$	$[stmt_2]^{-1}\ [stmt_1]^{-1}$
$\overset{pre}{\text{IF } condition}$ THEN $\quad\overset{if}{stmt}$ ELSE $\quad\overset{else}{stmt}$ $\overset{post}{\text{ENDIF } condition}$	$\overset{post}{\text{IF } condition}$ THEN $\quad\left[\overset{if}{stmt}\right]^{-1}$ ELSE $\quad\left[\overset{else}{stmt}\right]^{-1}$ $\overset{pre}{\text{ENDIF } condition}$
$\overset{from}{\text{FROM } condition}$ $\text{DO } \overset{do}{stmt} \text{ LOOP } \overset{loop}{stmt}$ $\overset{until}{\text{UNTIL } condition}$	$\overset{until}{\text{FROM } condition}$ $\text{DO } \left[\overset{do}{stmt}\right]^{-1} \text{ LOOP } \left[\overset{loop}{stmt}\right]^{-1}$ $\overset{from}{\text{UNTIL } condition}$
CALL *name*	UNCALL *name*
UNCALL *name*	CALL *name*
$\overset{1}{var} : \overset{2}{var}$	$\overset{1}{var} : \overset{2}{var}$
name += *expression*	*name* -= *expression*
name -= *expression*	*name* += *expression*
name ^= *expression*	*name* ^= *expression*
READ *name*	WRITE *name*
WRITE *name*	READ *name*

and its inverse. Either can be used as *forward*, and the other can then be invoked as its *reverse*.

8.3.3.7   Limitations of Features

Relative to modern software engineering features well developed in traditional forward-only languages, the following are some of the limitations of the **Janus** language:

■ All variables are global in scope. No specific concept of *local* variables is available, although a distinction of local versus global variables is widely used in all modern languages. In general, no concept of *scope* or *visibility* is defined.

■ All variables are of integer type. No other data type (such as floating point or character string) is supported. While this simplifies the language with respect to completeness and reversibility, additional types are needed for more practical purposes in many applications.

■ All variables are initialized to zero by the system. While modern forward-only languages support initialization constructs that permit different variables to begin with different intial values, the same in **Janus** must be explicitly added as an additional executable statement.

■ There is no concept of procedures accepting *formal arguments*, and no notion of passing actual arguments from the caller to callee. All data exchange with procedures is via global variables alone.

■ There is no concept of procedures *returning* values (i.e., procedures are subroutines but not functions). All side effects of procedures are achieved via modifications to global variables.

Nevertheless, these are not serious limitations, and can be remedied for software engineering purposes without affecting reversibility.

8.3.3.8   Example Programs

The original language definition of the **Janus** provides illustrations of how the language primitives can be used: a reversible factorization program, a reversible fibonnaci sequence generator, a reversible sorting program based on bubble sort, and a program to reversibly compute the square root of a number. Some interesting idioms that have more general applicability are

■ A method for reversibly doubling, by building only over the constructive operators += and -=,

■ A method in which, in a forward execution of a procedure, uses the reverse execution of another procedure.

8.3.4 R Reversible Language

The **R**, proposed as part of Michael Frank's doctoral dissertation [Frank, 1999], was originally intended as a reversible programming language for the Pendulum chip architecture [Vieri, 1995, 1999]. The Pendulum hardware supported the reversible Pendulum Instruction Set Architecture (PISA) aimed at *asymptotically zero-energy* computing [Younis and Knight, 1994, Vieri et al., 1998]. The **R** provided the high-level language interface to develop programs with PISA as the target for execution, and relied on the reversibility of PISA primitives for its own reversibility. The **R** incorporates several reversible constructs that are analogous to those of the **Janus** language, although both languages were independently developed and they differ in some aspects.

8.3.4.1 Control Flow Constructs

An **R** program consists of a main routine and zero or more subroutines that can accept formal arguments. Control flow is obtained via compositions of subroutine invocations, conditional statements, looping statements, and elemental statements.

- Subroutines may be invoked in forward mode using the syntax

$$(\text{ call } \textit{name} \text{ } [\textit{args}] \text{ })$$

  or in reverse mode using the syntax

$$(\text{ rcall } \textit{name} \text{ } [\textit{args}] \text{ }).$$

- Conditional statements with the syntax

$$(\text{ if } \textit{condition} \text{ } \textit{statement} \text{ })$$

  require the condition value to be identical before and after the body of the condition is executed.

- Iteration is supported with the syntax

$$(\text{ for } \textit{var} = \textit{start} \text{ to } \textit{end} \text{ } \textit{statements} \text{ })$$

  such that modifications to *start* and *end* are disallowed in the statements of the loop body, but modifications to *var* are permitted inside the statements of the loop body to change the number of iterations.

8.3.4.2 Data and Elementary Operations

All variables are of type integer, but can be scalars or arrays. Nested scoping of variables is provided via distinction between global and local variables.

■ Global scalar variables are declared as

$$(\text{ defword } name \ value \ )$$

while global arrays are declared as

$$(\text{ defarray } name \ value \ [\cdots \ value] \ )$$

to declare and initialize an array with the given number of values. References to array elements are made as *array_index*.

■ Local scoping is obtained with the syntax

$$(\text{ let } (var \ \texttt{<-} \ value) \ statements \ )$$

such that the variable *var* is valid for use within the given statements, with the requirement that *var* holds the same *value* at the end of the statements (e.g., if it was initialized to zero, it ends up with the value zero at the end of the scope).

■ Variables may be modified via the constructive operators: increment `++`, rotate in place left `<=<` and right `>=>`, add `+=`, subtract `-=`, and exclusive OR `^=`.

■ The swap operation

$$(\ var1 \ \texttt{<->} \ var2 \ )$$

exchanges the contents of two variables.

■ Expressions are composed using conventional recursive uses of terms and factors over unary and binary operators such as `+`, `-`, `&`, `<<`, and `>>`.

■ There is no specific input mechanism provided as part of the original language. Output is obtained via `printword` and `println` statements.

■ Reversible arithmetic is provided in terms of the increment and decrement operators mentioned before, in addition to a special multiplication operator of the form

$$(\ integer \ \texttt{*/} \ fraction \ ),$$

where *integer* i is a 32-bit integer, *fraction* f is 32-bit value interpreted as a fixed-point representation $f \in [-1, 1]$, and the result is $i \times f$, which is the same as the most significant 32-bits of the 64-bit integer obtained by the product of integer i and fraction f when f is viewed as an integer.

The semantics of this operator can be illustrated with a simple example using 4-bit numbers (instead of 32-bit numbers). Let a 4-bit integer variable I contain the value of 12, and another 4-bit integer F contain the value 8. Then, the value of F viewed as the fraction is $\frac{8}{2^4} = 0.5$. The special multiplication operator is evaluated as

$$I \ */ \ F \ = \ 12*0.5 \ = \ 6.$$

This is the same as the value of the leading 4 bits of the 8-bit integer obtained from the integer product I*F=96=16*6.

8.4   Functional and Logic Languages

Two principal characteristics of (pure) functional languages are the ability to pass functions as arguments to other functions, and the side effect-free operation of all functions. It is the latter that has interesting application to reversible execution. Essentially, in pure functional programming, the concept of assignment to variables does not exist. No variable can lose its value as there is no concept of overwriting the value of a variable. Consequently, the irreversibility-related problem of destructive assignment does not arise. Instead of overwriting variables, every computation generates a new set of variables, effectively creating a history of execution in memory. This aspect can be capitalized to realize reversible execution. To reverse any specific part of a computation, the results of that computation can be simply discarded because the previous values remain intact (because computation is side effect-free) and, unlike procedural languages, no additional step for restoration of values is needed. Logic programming also shares these features in the sense that the programming model does not have a concept of assignment to variables per se. These aspects have been exploited and reversibilty explored in different languages [Glück and Kawabe, 2003, 2004, Kawabe and Glück, 2005, Yokoyama et al., 2012].

However, in practice, for efficiency or practical purposes, in many functional languages, the side effect-free mode of execution is relaxed to minimize the memory used by the program. Memory for some variables is reused dynamically wherever the runtime system determines it to be possible without affecting the forward execution correctness. Values are destroyed via memory reclamation during execution whenever the runtime system perfoms garbage collection or prunes the history to free up some memory. Such optimizations violate the purity of side effect-free operation, which in turn interferes with the reversibility of execution because of the potential for loss of information.

Several functional languages have been proposed for reversible execution. The **INTERLISP** Language and environment [Teitelman, 1975, 1984] was an early LISP system that provided support for an extensive and powerful UNDO operation. The Ψ-LISP Language (or with the palindromic name **PSILISP**) incorporated reversible execution despite reuse of memory from garbage collection [Baker, 1992]. Baker also provided an interesting analogy between thermodynamic concepts such as temperature and software concepts such as number of live objects in a dynamic memory setting [Baker, 1992, 1994]. New constructs for reversible arithmetic are defined by expanding the semantics of the operators, such as reversible multiplication and division. The **pGCL** language [Zuliani, 2001] is a *probabilistic guarded-command language* that is amenable to automatic generation of reversible execution. The **Fun**, **Inv** and **LRInv** Languages are other examples of reversible functional or imperative languages [Mu et al., 2004, Glück and Kawabe, 2003, Mu et al., 2004, Kawabe and Glück, 2005].

8.5   Further Reading

For traditional (irreversible) languages, such as **C**, the compiler itself can be written in the same language. With the introduction of the new dimension called reversibility, an interesting exercise in completeness of reversible computation would be to write a compiler or interpreter of a reversible language as a program in the same language. In fact, this feat has been demonstrated for the **Janus** language: a self-interpreter called **SINT** for **Janus** programs is developed in **Janus** itself [Yokoyama and Glück, 2007].

Targets for reversible language programs could be traditional (irreversible) or reversible computing devices. Translation of programs in languages such as **Janus** can be easily performed for traditional device targets, because the forward and reverse semantics for every statement type can be easily mapped to a traditional language (but, obviously, not the other way around). However, for reversible devices, all statements can also be translated except for the evaluation of right hand side expressions in an assignment statement. Any complex expression will require a sequence of arithmetic and logical instructions in order to arrive at the final value of the expression before that value is used in the reversible assignment. This sequence of instructions need to be *reversibly* performed, which will require additional translation complexity. A solution is to utilize algorithms such as the one Charles Bennett proposed for simulating irreversible computation over a Reversible Turing Machine, as described previously in Section 6.5. Additional details on this topic are discussed in [Frank, 1999, Yokoyama and Glück, 2007].

Chapter 9

Adding Reversibility to Irreversible Programs

9.1   Overview

Because most programs are irreversible, reversibility must be added before
they can be used in a reversible computing context. Broadly speaking, there

are two major approaches to adding reversibility to irreversible programs: checkpointing and reverse computation. Checkpointing takes snapshots of the *memory* while reverse computation takes snapshots of the *control flow* of the program. More generally, checkpointing can in fact be subsumed by reverse computation when the unit of control flow is relaxed to vary in the spectrum of reversible units from simple arithmetic operations all the way to large subroutines. When an entire code fragment, such as a subroutine, is viewed as a single operation, the entire state potentially modified by the fragment is saved as control flow information, which essentially corresponds to checkpointing of the same data. However, it is useful to keep the distinction between checkpointing and reverse computation because of the optimizations that are possible to be performed more easily for checkpointing schemes when the unit of computation is known/assumed to be sufficiently large. In a unified, composite approach, there is little in the way of theoretical feasibility to have a single tool chain that subsumes all the different approaches. The range of alternatives is shown in Figure 9.1, each of which is described next.

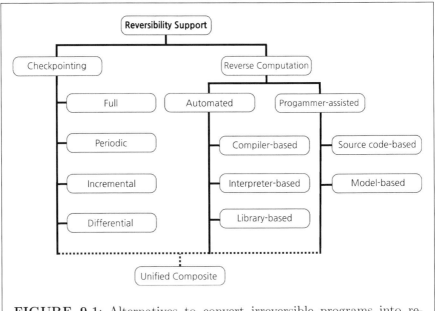

FIGURE 9.1: Alternatives to convert irreversible programs into reversible ones.

9.2   Checkpointing

Reversibility can be added to any irreversible program by taking snapshots of the program state prior to every modification to the state. All types of modifications can be divided into units called actions. For example, with fine-grained units, each subroutine call or each executable statement can be considered an action. Coarser-grained units may involve entire subroutines. Checkpointing is the process of taking snapshots prior to actions so that the system can be restored to its state prior to any specific action. Different checkpointing schemes are available that give different snapshot frequencies and varying snapshot sizes.

Here, four major checkpointing schemes are presented. The first is *Full Checkpointing*, which is the simplest of the schemes. The second is *Periodic Checkpointing*, which is a relaxation of *Full Checkpointing* to provide a lever for controlling the frequency of checkpointing to reduce the overall checkpointing overhead. The third is *Incremental Checkpointing*, which works well when the state to be checkpointed is large but modifications are sparse and scattered. The fourth is *Differential Checkpointing*, which also works well for sparse and scattered modifications to a large state but trades off additional storage in return for increased speed of reversal relative to *Incremental Checkpointing*.

Each of the four schemes can be viewed as a basic checkpointing building block, and more complex composite schemes can be employed as a combination of the basic building blocks. For example, it is possible to combine incremental checkpointing with periodic checkpointing using incremental checkpoints between consecutive pairs of periodic checkpoints. Similarly, full checkpoints can be used to reset the differential checkpoints whenever the size of the differential checkpoint increases significantly.

9.2.1   Full Checkpointing

In this style of checkpointing, the entire state of the system is saved after every action. This is most applicable when one or more of the following conditions apply: (1) it is difficult to identify or delineate which portions of the state have been modified by the action (2) the total size of the system state is so small that the memory cost and/or runtime cost of making a copy of the state are insignificant compared to the temporary storage and/or the computation time taken by the action, or (3) the storage is easier to organize and manipulate in terms of fixed and unchanging size of the state. Full checkpointing, in most cases, represents a brute-force approach to reversibility and seldom the most efficient, although there are some cases where there is no better recourse. The logical view of full snapshots at every action, however, serves in general as a useful basis for terminology and a baseline for runtime efficiency.

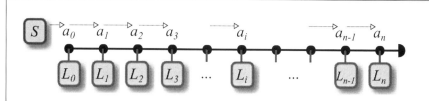

FIGURE 9.2: Full checkpointing for snapshots-based reversal.

9.2.1.1 Operation

The schematic of full checkpointing is shown in Figure 9.2. The state of the original irreversible system is the set of variables $S = \{s_j | 0 \leq j < N\}$. The sequence of actions $A = (a_0, \ldots, a_n)$ performs a sequence of modifications to subsets of variables in S. To add reversibility to the actions, a copy of the entire state is made before every action. Thus, a full copy L_i of S is made before action a_i; the copy holds the value of all the state variables before the action is performed. Later, if and when a_i needs to be reversed (or, all $a_j, i \leq j$, need to be reversed), it is sufficient to copy L_i onto S to restore the state to the correct value it had before a_i was previously executed.

9.2.1.2 Implementation

Full checkpointing is one of the easier schemes to implement. The copies are maintained in a stack of fixed-size buffers, as reversal is always in last-in-first-out order. Just before any action a_i is executed, a buffer L_i is allocated and the state S is copied into it. A *deep copy* operation is necessary because actions are assumed to operate on any part of the memory, including by following pointers to jump across memory locations. Because a deep copy is in general difficult to implement (and sometimes expensive in terms of runtime cost, to account for cycles created via pointers), it is beneficial to organize the state S as a contiguous sequence of bytes that are copied using byte-copying primitives.

9.2.2 Periodic Checkpointing

Periodic checkpointing is essentially the same as full checkpointing, with one major difference, namely that a full checkpoint of the state S is saved to a copy P_i for every p actions, instead of for every single action.

9.2.2.1 Operation

The schematic of periodic checkpointing is shown in Figure 9.3. Full checkpointing can be viewed as a special case of periodic checkpointing, with $p = 1$. Restoration to a point prior to any specific action a_{mp} is a simple matter

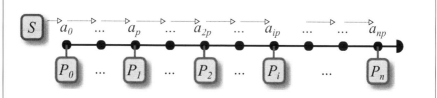

FIGURE 9.3: Periodic checkpointing for snapshots-based reversal.

of restoring the checkpoint P_m to S. To reverse to any intermediate action a_i that does not have a corresponding checkpoint (i.e., $i \neq p \left\lfloor \frac{i}{p} \right\rfloor$), the state must be recreated by first restoring to the latest checkpoint that precedes the intermediate checkpoint. This re-creation is achieved by re-executing the actions from the preceding checkpoint to the point of reversal. All actions $a_{pk}, a_{pk+1}, \ldots$ are reset, where $k = \left\lfloor \frac{i}{p} \right\rfloor$. The checkpoint P_k is then copied into S, followed by re-execution of actions $a_{pk}, \ldots, a_{i-1}$.

9.2.2.2 Implementation

The implementation of periodic checkpointing is similar to that for full checkpointing, with some additional bookkeeping to record which actions have a checkpoint and which do not. Reversal is similarly modified to search backward for the most recently available checkpoint while traversing the list of actions in the reverse direction.

9.2.3 Incremental Checkpointing

In some programs, actions do not update the entire state, but instead modify a few scattered items of the state. In such cases, it is sufficient to make a copy of only those variables that are about to be modified, rather than save the entire state of the program. Thus, the state is *incrementally* saved over execution.

9.2.3.1 Operation

The schematic of incremental checkpointing is shown in Figure 9.4. Each action a_i results in a pre-modification copy δ_i of set of the modified variables *relative to the most recent modification* to the state. In comparison to full checkpointing, the snapshot of the system before action a_i is obtained by accumulating the set of incremental changes from the beginning to that action. Thus, $L_i = S \oplus \delta_0 \oplus \delta_1 \oplus \ldots \oplus \delta_i$, where the operation $S \oplus \delta$ indicates applying a change set δ to S.

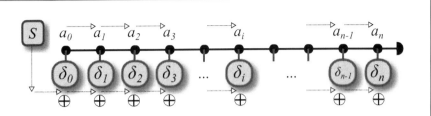

FIGURE 9.4: Incremental checkpointing for snapshots-based reversal

9.2.3.2 Implementation

Among the checkpointing schemes, incremental checkpointing is in general the most efficient scheme, but is also one of the more challenging ones to implement. The basic problem that poses the primary challenge is the identification of the specific portions of the state that is changed by the actions *before* those portions are modified. In small programs, it is possible to identify every update to every variable and mark them to signal a modification so that the values can be saved before they are overwritten. However, in programs of nontrivial size, several complications arise. Without compiler help, it is very difficult to trap all modifications to every variable. Also, pointers to variables obfuscate variable identities, making it difficult to detect the first modification to any variable.

A compiler-based approach can be used to instrument the generated code to make a copy of every variable upon first modification to their value as part of an action, although even with a compiler, it is relatively difficult to find the first modification to array elements without some additional memory overhead for keeping modification flags (bits) for each element.

Without compiler support, there are a few ways to trap modifications to the variables. One approach is to rely on the programmer to manually indicate within the program the points of first modification to variables within any action. This is achieved by the programmer inserting an invocation to an incremental checkpointing routine prior to modification in the source code. Another approach is to use operator overloading in object-oriented languages such as C++. Inside the overloaded methods, the first modification is logged to incremental storage stack and then the operation is applied to the variable. For example, a declaration such as `int x;` for an integer variable named x is replaced by `Integer x;` where the type `Integer` is a wrapper object around the primitive integer type with overloaded operators for assignment such as =, +=, -=, and so on.

The operator overloading approach facilitates some amount of automation; however, the programmer is still required to alter the declaration of the object types to the new wrapper types (e.g., `int` to `Integer`). Another approach to incremental checkpointing is the *binary editing* approach in which the object code in the binary executable is processed to automatically identify and insert instrument modifications to variables. The binary editing approach is an automated solution that can add incremental checkpointing capability after compilation, but it is also the least portable and least amenable to optimization due to lack of source code-level information.

In the underlying runtime support, incremental checkpointing is realized as a trace of change sets $\{\delta_0, \ldots, \delta_i, \ldots\}$ logged in a stack (last-in-first-out) data structure. Each change set is an encoding of the place in memory that is being modified, and the value that is being overwritten. For example, for every variable x that is about to be modified in forward mode, the pair $\delta = (\&x, \hat{x})$ is logged to the incremental checkpointing stack, where $\&x$ denotes the pointer to the variable x, and $\hat{x}$ denotes the value of x. In reverse mode, the trace is traversed backward, and each change set $\delta_i = (\&x_i, \hat{x}_i)$ is reversed by copying $\hat{x}_i$ to the location to which $\&x$ points.

When the size of each variable being saved is small, the overhead for incremental checkpointing can be large. For example, to incrementally checkpoint a modification to a 4-byte integer variable, assuming 64-bit pointers, 12 bytes are logged to the checkpoint storage, which constitutes a 200% overhead relative to the size of the variable itself. Thus, incremental checkpointing is more effective than full or periodic checkpointing only if the number of variables actually modified is much less than the total number of variables potentially modifiable in an action.

Another drawback of incremental checkpointing is that the state restoration cost for reversal is proportional to the execution length because the incremental trace must be sequentially traversed backward to the reversal point. Also, due to the sequential nature of traversal along the trace, the same variable may be restored multiple times (because the same variable may appear multiple times in the log), although it is sufficient to restore only the earliest copy. Because the earliest copy cannot be determined without traversing the trace, it is in general not possible to avoid multiple restorations. The sequential cost can be significant if the log is long as a result of a large number of variables being modified in the forward mode.

9.2.4 Differential Checkpointing

In programs that modify small, scattered portions of their state repeatedly over major lengths of execution, full and periodic checkpointing schemes are suboptimal due to the sparse nature of modifications, and incremental checkpointing is suboptimal due to large amount of duplication of snapshots for the same set of variables. In such cases, it is more efficient with respect to runtime and memory cost to save change sets relative to the original (initial) state instead of relative to previous action. This is achieved via differential checkpointing, which generates a log of change sets that are relatively larger in individual size but whose composite size is smaller than the aggregate size ot their equivalent incremental change set log; also differential checkpointing can restore the state to any action faster than via incremental checkpointing.

9.2.4.1 Operation

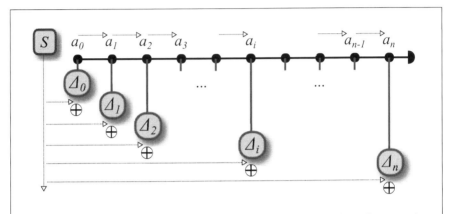

FIGURE 9.5: Differential checkpointing for snapshots-based reversal.

The schematic of differential checkpointing is shown in Figure 9.5. Each action a_i results in a pre-modification copy Δ_i of set of the modified variables *relative to the initial value* of the state. This is different from incremental checkpointing in that the snapshot of the system before action a_i is obtained by taking the difference from the initial state, as opposed to the value after the most recent action. Thus, $L_i = S \oplus \Delta_i$, where Δ_i is the set of differences between S and the value after action a_i is applied, and the operation $S \oplus \Delta$ indicates applying a change set Δ to S.

9.2.4.2 Implementation

Differential checkpointing can be implemented using two basic approaches. In both cases, a copy of the initial state S_0 is kept in addition to the evolving state S.

The first is the *dirty bit* approach in which every variable s_j is associated with a *dirty bit* d_j and a shadow value $\hat{s}_j$. The dirty bits of all variables are in unset status before any action. During any action a_i, whenever a state variable s_j is modified, its dirty bit d_j is set and a pre-modification copy is saved into its shadow value $\hat{s}_j$. At the end of every action a_i, all state variables whose dirty bits are set constitute the differentially checkpointed set Δ_i for that action; a copy of shadow values for all those dirtied variables is logged to the differential checkpoint log at the end of each action. Dirty bits are *not* reset at the end of actions—this is what enables the differential nature of the checkpoint with respect to the initial state. When this dirty bit scheme is used in combination with other schemes such as full or periodic checkpointing, all dirty bits must be reset to the unset status any time a full checkpoint is taken. This resetting of dirty bits effectively resets the differential checkpoint to null and prevents the differential checkpoint size from growing too large.

The second approach is the *computed differences* approach in which at the end of every action a_i, the difference Δ_i between the current state S and the initial state S_0 is computed ($\Delta_i = S - S_0$) and saved to the differential checkpoint log.

In both cases, reversal to any action a_i is achieved by first overwriting the current state S with the initial state S_0 and then incorporating the change set Δ_i onto S.

The dirty bit approach is useful when system support is available to detect changes to memory units. For example, several operating systems provide detection of modifications at the level of virtual memory pages, and user-specified handling of automatically generated notifications (such as via a signal). Changes to files in a file system are also available for detection via file status flags, in the form of an operating system-maintained *archive bit*. Advanced features such as copy-on-write semantics are also available in some operating systems that support shared and cloned pages across processes, which can be exploited for incremental or differential checkpointing schemes.

9.2.5 Example Application

The different checkpointing schemes can be seen in the following example. Consider a program in which the motion and collisions among N hard spheres in d dimensions are simulated reversibly [Perumalla and Protopopescu, 2013]. Each particle c_i is represented by its d-dimensional position vector x_i and d-dimensional velocity vector v_i. Thus, the system state consists of $S = (X, V)$, where $X = \{x_i | 0 \leq i < N\}$ and $V = \{v_i | 0 \leq i < N\}$. The simulation proceeds as a sequence of collisions; the modification of the system state at each collision is realized as an *action* that changes the values of some or all of the variables in the system state S. Thus, the system evolves as a sequence of actions $(a_0, \ldots, a_i, \ldots, a_n)$, and the checkpointing schemes ensure that the system is capable of being restored to its state before any action $a_i, 0 \leq i < n$.

With full checkpointing, because each collision creates an action that updates the state, the entire state (X, V) of all particles is saved upon every collision. If the simulation updates the positions of all particles, then the full copy of the state is indeed needed for reversal.

With periodic checkpointing, the positions and velocities are saved after every p collisions, where p is determined by system characteristics such as the ratio of computational speed to memory access times.

With incremental checkpointing, the positions and velocities of only the particles affected during a collision are saved prior to the collision. This works well when the number of collisions is large and almost all the particles are involved in some collision at some point in simulation, and the average number of collisions for any given particle is small.

In differential checkpointing, the positions and velocities of all the particles that have undergone collision since the initial state are saved after every collision. This works well when a relatively small set of particles undergo many collisions, and not every particle is involved in collisions.

9.3   Reverse Computation

In contrast to taking snapshots of the state, reversibility can be added to irreversible programs by *computing* in reverse. Reverse computation can be added to a program either via automated means, semi-automated means, or manual introduction by the programmer. Automated approaches include compiler-based reversal and interpretation-based reversal, and semi-automated approaches include library-based reversal. In manual approaches, the programmer can perform a source-code transformation to mimic how a compiler-based approach would operate, except that the programmer can use knowledge about data structures and algorithms used in the program that the compiler cannot glean by automated means. Manual approaches to reversal also involve using domain-specific knowledge for reversal that are not possible to automatically generate because the forward code and reverse code are not related by source code correspondence but are related by domain-specific insights and semantics.

9.3.1   Automated: Compiler-Based

The first idea is to use a compiler to parse the input (irreversible) program and somehow transform it into a reversible one. However, although conceptually straightforward, writing a working, general-purpose compiler to automatically add *efficient* reversibility to complex programs is an extremely challenging task. While a first-cut reversal is relatively easy for simple programming languages, complications quickly arise when more control flow constructs and complex data types are considered for automated reversibility. Thus, for advanced high-level programming languages, a compiler-based reversal is highly challenging.

Consider a simple irreversible code fragment given in Algorithm 9.1. The compiler-based approach generates two versions from the code, one for forward execution and the other for reverse execution, as given in Algorithm 9.2. Note that control flow information is tracked via two new variables introduced by the compiler: the variable c remembers the count of the number of iterations of the while loop, while b remembers one bit of information about which branch of the **if** statement was taken in the forward mode. Although the reversible

Algorithm 9.1 A simple irreversible program code fragment for adding reversibility

subroutine $f()$		
I_1, R_1, W_1		
while (R_{while})		
$\quad I_2, R_2, W_2$	I_i $\quad = \quad$	i^{th} non-control flow instruction
$\quad I_3, R_3, W_3$	I_i^{-1} $\quad = \quad$	Inverse instruction of I_i
end while	R_i $\quad = \quad$	Set of variables read by I_i
if (R_{if})	W_i $\quad = \quad$	Set of variables overwritten by I_i
$\quad I_4, R_4, W_4$	R_{while} $\quad = \quad$	Variables used in loop condition
else	R_{if} $\quad = \quad$	Variables used in branch condition
$\quad I_5, R_5, W_5$		
end if		
I_6, R_6, W_6		
end subroutine		

forward and reverse codes can be optimized to minimize or eliminate the space used to remember the control flow information, this example is kept simple to illustrate the high-level idea of compiler-based reversal. Note also that some instructions may need additional memory space to store some information that may be needed for their reversal; such needed additional variables are omitted for simplicity in the example.

Compiler-based approaches can be divided at a high level into two categories. One is the *source-to-source compilation* approach, and the other is the *augmented compiler* approach.

9.3.1.1 Source-to-Source Translator

In the source-to-source approach, the irreversible code is fed to a source-to-source translator that creates a new, reversible version of the code while retaining its forward semantics unchanged. Any given block of source code is statically transformed via compilation into a forward computation block and a corresponding reverse computation block. The forward computation block preserves the semantics of the input source code in addition to making it reversible. The reverse block then corresponds to its inverse. Both the forward and reverse blocks are then compiled just as normal code. This approach is illustrated in figure Figure 9.6 for a code fragment.

The translator operates entirely at the source code level and hence only contains transformations to the abstract syntax tree (AST). The transformations include insertion of new source code for instrumentation of control flow, and generation of new reverse peers of the input (forward) components (e.g., reverse versions of forward subroutines). Optimizations to minimize or eliminate instrumentation in the forward mode are also performed by the translator. This approach is illustrated in Figure 9.7. The translator gener-

Algorithm 9.2 Example of reversal using the compiler-based approach

Forward	Reverse
subroutine $f()$	**subroutine** $f^{-1}()$
$\quad I_1, R_1, W_1$	$\quad I_6^{-1}, R_6, W_6$
$\quad \boxed{c \leftarrow 0}$	$\quad$ **if** $(b = 1)$
$\quad$ **while** (R_{while})	$\quad\quad I_4^{-1}, R_4, W_4$
$\quad\quad \boxed{c \leftarrow c + 1}$	$\quad$ **else**
$\quad\quad I_2, R_2, W_2$	$\quad\quad I_5^{-1}, R_5, W_5$
$\quad\quad I_3, R_3, W_3$	$\quad$ **end if**
$\quad$ **end while**	$\quad$ **while** $(c > 0)$
$\quad$ **if** (R_{if})	$\quad\quad c \leftarrow c - 1$
$\quad\quad \boxed{b \leftarrow 1}$	$\quad\quad I_3^{-1}, R_3, W_3$
$\quad\quad I_4, R_4, W_4$	$\quad\quad I_2^{-1}, R_2, W_2$
$\quad$ **else**	$\quad$ **end while**
$\quad\quad \boxed{b \leftarrow 0}$	$\quad I_1^{-1}, R_1, W_1$
$\quad\quad I_5, R_5, W_5$	**end subroutine**
$\quad$ **end if**	
$\quad I_6, R_6, W_6$	
end subroutine	

ates the forward and reverse versions of the code, which are fed to a traditional forward-mode compiler to generate the final executable that is reversible. The traditional compiler views the reverse code as a forward version and compiles normally as such, oblivious to the fact that the reverse code is designed to undo a prior invocation of the corresponding forward code. Typically, only the parts of the program that need reversal are fed to the translator, while the rest of the application is compiled normally (without passing through the translator) and linked to the final reversible executable.

A major advantage of the source-to-source compilation approach to reversibility is that it provides excellent portability while requiring no changes to existing software infrastructure such as native compilers, linkers, and loaders. Moreover, all the decades of advanced compilation technology can be brought to bear on the reverse code as well, without any additional effort, thus easily optimizing the reverse mode code for system-level effects such as better cache performance. A difficulty with this approach is that an entirely new translator needs to be written for every programming language that needs to be reversed, which would involve duplication of major portions of existing compiler infrastructures and frameworks.

9.3.1.2 Augmented Compiler

In the augmented compiler approach, the traditional forward-mode compiler is augmented with reversible execution capabilities as part of its core

if(qlen < B) {	if(qlen < B) {	if(b == 1) {

`  dclays[qlen]++;`
`  qlen++;`
`} else {`
`  lost++;`
`}`
	`  b = 1;`
	`  delays[qlen]++;`
	`  qlen++;`
	`} else {`
	`  b = 0;`
	`  lost++;`
	`}`
		`  --qlen;`
		`  --delays[qlen];`
		`} else {`
		`  --lost;`
		`}`

Original	Forward	Reverse

FIGURE 9.6: Example of the compiled approach to adding reversibility.

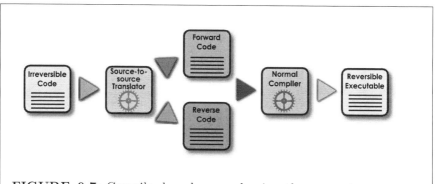

FIGURE 9.7: Compiler-based reversal using the source-to-source method.

functionality. Generation of reversible code, as opposed to the traditional forward-only code, would be a compiler-provided feature invoked via, say, command-line options of the compiler. This is analogous to functionalities such as generation of thread-safe code or position-independent code commonly provided as compiler options to specify the type of code emitted as output by the compiler.

The methodology of the augmented compiler approach is shown in Figure 9.8. The irreversible program is provided as usual to the compiler. The compiler internally generates forward code *augmented* for reversal, and also synthesizes reverse code that relies on the runtime information generated from the augmentation of the forward code. Object codes for both forward and reverse codes are generated, which are used to form the reversible executable.

The augmented compiler approach provides the advantage of transparent support for reversibility. However, it is the least portable approach because

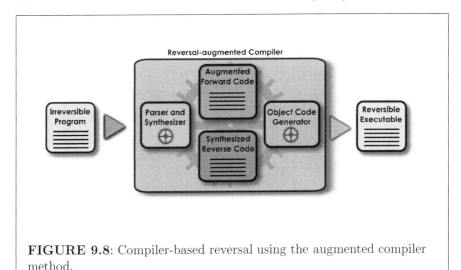

FIGURE 9.8: Compiler-based reversal using the augmented compiler method.

it requires modification of the internals of every compiler, which may be impractical in the short term.

9.3.1.3 Memory and Time Costs

With the compiler-based approach, memory is used in the instrumentation of control flow, and runtime is increased due to the instrumentation code. Forward code size is only increased by a constant factor because each original instruction (in the forward-only code) is instrumented using an addition of a constant number of instructions (typically, a single additional instruction). Thus, the code size complexity remains $O(1)$ with respect to execution length because the code remains constant, independent of the run length. However, additional variables introduced for instrumentation of control flow require additional memory for storage, which increases the overall data size. For example, a counter variable introduced to remember the number of iterations executed by a variable-iteration loop requires $O(\log L)$ bits, where L is the execution length in terms of the number of instructions. A branch instruction within a loop can introduce the need to store up to L different outcomes of the branch instruction, in which case the data size becomes $O(L)$ bits. The overall runtime of forward execution remains $O(L)$ because the number of instructions in the augmented forward mode is proportional to the original execution length.

9.3.2 Automated: Interpreter-Based

In the interpreted method for reversal, the forward computation code is augmented with special instrumentation code. During the actual forward execution, the specially added code generates a sequential log of all the individual operations that are executed. For reversal, the generated log is then interpreted in the reverse order and appropriate inverse operations are invoked. This approach is illustrated in Figure 9.9 for a code fragment. Typically, the operation indicators along with actual operand values are saved during forward computation, which are later interpreted during reverse execution, and their corresponding inverse operations are applied to the operands.

| ```
if( qlen < B ) {
  delays[qlen]++;
  qlen++;
} else {
  lost++;
}
``` | ```
if( qlen < B ) {
  LOG(INCR,
    &delays[qlen]);
  delays[qlen]++;
  LOG(INCR,&qlen);
  qlen++;
} else {
  LOG(INCR,&lost);
  lost++;
}
``` | ```
for all log entries
in reverse order:
switch( operation )
{
  case INCR:
  {
    (*p)--;
  }
  ...
}
``` |
| Original | Forward | Reverse |

FIGURE 9.9: Example of the interpreted approach to adding reversibility.

Interpreter-based reversal is one of the easier approaches to implementing reversible execution. The program itself requires no modification, and all the reversibility support can be added entirely within the interpreter engine. When irreversible programs written in an (irreversible) interpreted language are executed by the interpreter engine, reversibility for the program can be enabled by modifying the interpreter engine to generate three distinct tapes at runtime, as shown in Figure 9.10. The first is the *output tape* that generates the normal output of the original forward-only execution of the program. This output is augmented with reversibility such that output can be consumed as input when the interpreter is executed in reverse mode. In addition to the normal output logged on the output tape, two new tapes are generated. The first is the *instruction tape* to which the interpreter writes the sequence of primitive instructions. These are instructions that do not contain any conditional or looping constructs, so that the sequence is entirely linear in nature. The second is the *data tape* that contains all the data (operands of instructions) that are required for the reversal of the linear sequence of instructions logged on the instruction tape. A flag is provided in the interpreter that the program user can use to switch the mode of execution from forward to reverse or vice versa, dynamically at runtime.

For the simple irreversible code given in Algorithm 9.1, the interpreter-
```

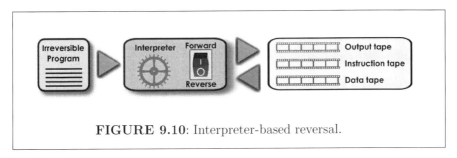

FIGURE 9.10: Interpreter-based reversal.

based approach executes the instruction tape in forward and backward modes as given in Algorithm 9.3. Note that no control flow information is present in the instruction tape, such as branch or jump statements; instead, only a sequence of instructions that directly update the variables are executed and logged. For example, there are C copies of the body of the while loop logged to the instruction trace, assuming C iterations of the loop. Note also that the read sets and write sets may vary across each instance of the loop body (e.g., due to array indices varying across loop iterations).

Algorithm 9.3 Example of reversal using the interpreter-based approach

Forward	Reverse
I_1, R_1, W_1	I_6^{-1}, R_6, W_6
I_2, R_{2_1}, W_{2_1}	I_5^{-1}, R_5, W_5
I_3, R_{3_1}, W_{3_1}	or
I_2, R_{2_2}, W_{2_2}	I_4^{-1}, R_4, W_4
I_3, R_{3_2}, W_{3_2}	$I_3^{-1}, R_{3_C}, W_{3_C}$
$\vdots$	$I_2^{-1}, R_{2_C}, W_{2_C}$
I_2, R_{2_C}, W_{2_C}	$\vdots$
I_3, R_{3_C}, W_{3_C}	$I_3^{-1}, R_{3_2}, W_{3_2}$
I_4, R_4, W_4	$I_2^{-1}, R_{2_2}, W_{2_2}$
or	$I_3^{-1}, R_{3_1}, W_{3_1}$
I_5, R_5, W_5	$I_2^{-1}, R_{2_1}, W_{2_1}$
I_6, R_6, W_6	I_1^{-1}, R_1, W_1

In a variant of this interpreter-based approach (e.g., [Biswas and Mall, 1999]), instead of storing the *forward* instructions, the inverses are written on the *instruction tape*. This moves some of the cost of reversal to the forward path, but removes that cost from the reverse path.

The ease of implementation of the interpreter-based approach is due to the fact that the interpreter engine contains all the information needed to *transparently* record all the information necessary at runtime. Reversal is also greatly facilitated by the fact that the logged instructions are already flattened into a serial trace of operations, with no branching or looping. While this is an advantage, it is also a shortcoming of the approach because it loses higher-level semantics (e.g, loop structure) that could be useful for effecting optimizations.

9.3.2.1 Memory and Time Costs

Consider an execution of L instructions of the irreversible program. The original code size is unchanged with interpreter-based reversal, and remains $O(1)$, being independent of the execution length. However, a memory cost of $O(L)$ is incurred for the instruction tape, as every operation is logged to the instruction tape. A similar worst-case memory cost of $O(L)$ is incurred to log the operands to the data tape.

9.3.3 Automated: Library-Based

The library could be built from code that is generated via compiler-based or interpreter-based reversal, or even from manually generated reversible code. Invocations to reversible libraries can be automatically generated by the compiler or can be manually introduced by the programmer. One of the best examples of this type of reversibility support is a library for reversible random number generation (see Chapter 12). With the library-based approach, the programmer is shielded, for the most part, from the intricacies of reversal. For a call to a subroutine $f()$ in a library, for example, the user would only need to invoke its inverse $f^{-1}()$ supplied with the same library. Such reversible libraries may be supplied by the operating system or by software vendors, similar to the conventional set of libraries shipped with most computing systems. This approach provides the least effort path for the programmer community, but transfers most of the burden of development effort and adoption to the system vendors.

9.3.4 Programmer-Assisted: Source Code-Based

In complex programs, it is often the case that the programmer has much more knowledge about data dependencies and evolution behavior than can be gleaned by the compiler. In one of the common situations, the compiler can find it hard to discern if an assignment or a group of apparent modifications to a variable are in fact irreversible. Yet, a reversal for the same could be discerned by the programmer. For example, a swap operation as a sequence of three assignments t=a; a=b; b=t can be mistaken by a naïve compiler as three destructive updates that need memory equal to the size of three variables; however, the same operation is perfectly reversible with no additional memory (by applying the same code again on a and b). Overall, the programmer can undertake the task of developing the reversal, but with some compiler assistance restricted only for the tedious, mechanical tasks. The source-to-source compiler approach may be used for the most part, but key programmer-provided information may be given to the compiler to more efficiently reverse certain portions of the code. For example, the compiler may generate reverse code by mechanical processes; however, the programmer annotates the original code with directives to the compiler via pragmas, say, to indicate that a branching condition is unchanged within either body of the branch statement, or that a loop's iteration count is unchanged within the loop body.

9.3.5 Programmer-Assisted: Model-Based

In the most sophisticated use of reverse computation, the reverse code may in fact be generated not from source code-based approach to reversal, but may originate from a theoretical reverse model corresponding to the forward model. For example, in a scientific simulation of particles undergoing elastic collisions, the collision operator that transforms the velocities of colliding particles may be mathematically complex. However, an exact inverse operator may be available as a separate model that can be coded and incorporated into the program for reversal. Moreover, such an inverse operator cannot be obtained automatically by any compiler-based technique because the reversal requires a great amount of domain-specific expertise. Another example is the reversal of linear algebra operations that are very hard to automate but easier to develop using theoretically well-known identities and relations in linear algebra. It is in fact these types of codes for which *reversible languages* are required (Chapter 8) so that the modeler can express the inverse codes using reversible constructs to retain the reversibility properties about which the modeler is aware.

9.4 Unified Composite

At a high-level, checkpointing might appear to be a distinct method different from reverse computing. Checkpointing schemes achieve reversal by tracking the changes to the data while reverse computation schemes achieve reversal by tracking the changes to the control flow. Nevertheless, checkpointing is theoretically a proper subset of reverse computation because every checkpointing operation can be viewed as a reverse computation item in which the unit of checkpoint is a destructive update to a correspondingly sized unit of memory.

Consider a series of updates to a set of data variables $D = \{d_1, \ldots, d_n\}$ performed as part of a sequence of actions $A(D) = (a_1, \ldots, a_m)$. Checkpointing performed on D can be achieved via reverse computation by viewing A as a single destructive update to a single datum D, and this aggregate action A is included as one of the operations to be reversed in the overall reverse computation framework. Thus, it is conceivable to treat checkpointing as a natural outcome of a broader, optimized reverse computation framework in which information logged to the history trace is automatically minimized by choosing the smaller of the modified data size and the control flow information.

In the preceding example, suppose that the bit size of the data being modified is $M_D = |D|$ and the number of bits needed to reverse the actions is $M_A = |A|$. If $M_D < M_A$, it implies that the actions perform several irreversible operations on the data, and the reversal of each of those actions consumes more bits of memory than the size of the modified data itself. In this case, the entire set of actions A can be abstracted as a single destructive update to a single datum D, and checkpointing can be used to save the data. This becomes a single operation in the reverse computation view, which consumes only M_D bits for reversal. On the other hand, if $M_A < M_D$, then, memory-wise it is more economical to save the M_A bits of control flow information that can be used to uncompute for reversal.

For example, consider adding reversibility to a sequence of N invocations made to a pseudorandom number generator. There is a trade-off point $N > N^*$ after which checkpointing is more economical than reverse computation. When $N \leq N^*$, it may be more effective to use reverse computation to compute backward to restore the random number seed. However, as N increases beyond N^*, the runtime cost becomes larger than the gain from avoiding the memory copying operation. In that case, checkpointing works better. The determination of N^* varies with the actual computation and memory cost ratios. In general, for different codes and hardware systems, it is possible to give different weights to memory and computation and choose the best mix and trade-off among various schemes to arrive at a composite scheme for reversal.

Similarly, the various techniques within reverse computation, such as compiler-based, interpreter-based, and manual methods, can all be combined into a unified composite that subsumes the individual solutions applied to different portions of a larger codebase. For example, a reverse compiler can make use of optimized reversible libraries, and programs can provide user-defined mappings of forward-reverse pairs of functions with user-supplied code. The compiler and/or user can utilize a compilation-based approach or an interpreter-based approach to implement any given reversible function. For example, an interpreter-based approach may be used for a function within a reversible library linked to a code compiled using a reversible compiler-based framework. User-specified model-based reverse codes can also be similarly incorporated into such a framework. A grand, unified composite approach is thus conceivable in the long term to subsume all the methods to add reversibility to irreversible programs.

9.5 Further Reading

Many of the issues and considerations described in this chapter can be appreciated in the context of computing applications, such as automatic differentiation (or algorithmic differentiation) [Griewank and Walther, 2008] and parallel discrete event simulation [Fujimoto, 2000], in which the variety of needs in reversibility of computation gives rise to a corresponding variety in choosing the most appropriate reversal implementation.

The Reverse Mode of adjoint computation in Automatic Differentiation, for example, can be implemented in any of the aforementioned ways of reversal, each giving a different mix of computational and memory usage behavior. When checkpointing is used, the same variable may be stored repeatedly if the variable appears on the right-hand side of more than one variable, due to which incremental checkpointing becomes inefficient. A differential checkpointing scheme is employed in such cases, using a status flag per variable to indicate if the variable is "to-be-recorded." Reverse computation (also referred to as inverse computation in Automatic Differentiation literature) is employed to rematerialize a lost value based on inversion of a symbolic expression over a variable whose value is available.

Almost all the reversal techniques described in this chapter have been attempted in implementing parallel discrete event simulation systems [Lin and Preiss, 1991, Fujimoto et al., 1992, Fujimoto, 2000]. In the so-called optimistic style of parallel discrete event simulation (e.g., the Time Warp algorithm [Jefferson, 1985]), event computations require reversibility because they are subject to rollback. This reversibility of event computation can be obtained using any of the techniques described in this chapter. The problem of high overhead due to full checkpointing formed a stumbling block in early discrete event simulation engines [Jefferson et al., 1987, Hontalas et al., 1989]. Incremental state saving and differential checkpointing lessened part of the overhead when applied to complex simulation models [Perumalla et al., 1998]. The overheads were later significantly relieved using reverse computation techniques [Carothers et al., 1999, Perumalla, 2007].

Chapter 10

Reverse C Compiler

10.1 Reversibility of C Language Programs

The **C** language is one of the most successful and popular languages, enjoying the base of an extremely large, worldwide programmer community. However, it is an irreversible language, containing many irreversible constructs at the outset. Adding reversibility in a transparent or semi-transparent fashion to the execution of **C** programs can provide the immense benefit of bringing reversible computing to a large base of the existing programming community, and can also help stir additional research in incrementally enhancing the efficiency and applicability of reversible **C** programs.

Programs written in popular general-purpose languages tend to be irreversible in nature. Due to the very nature of the language constructs, some information is destroyed as the program executes, making the program impossible to execute in reverse. However, any program can be made reversible by carefully augmenting it in such a way that the information being destroyed by the program is actually saved on the side using additional nonintrusive instructions. The augmented program thus not only retains the semantics of the original program, but also becomes reversible. Such an augmenting approach is described here, which can be used to transform any given (irreversible) program into a semantically equivalent reversible program. The automatic generation of the reverse program is also described. Such automated means for reversibility, as presented here, is useful in application codes too complex for manual transcription and optimization by human inspection. Although the techniques are described for the **C** language, the approach is equally applicable to other procedural languages. The methodology described here has the following primary objectives:

- For **C** programs that do not contain any irreversible constructs, the compilation should produce a reversible execution whose memory needs do not grow with execution length. This can be achieved in an incremental manner by starting with a small-sized reversible subset of the language, and slowly expanding the set to include more reversible constructs. For example, an initial subset can contain the well-known operators such as increment and decrement operators, fixed-iteration loops, conditional statements in which the predicate expression is not modified in the conditional statement bodies, and so on. More sophisticated situations can be added to the memory-less reversibility, such as tracking the global data flow to overcome apparently destructive updates that in reality only shift the information around among variables (simplest examples of this being swap or circular rotation operations).

- For **C** programs that contain irreversible constructs, the compilation framework should provide optimization methods to minimize the overall cost for a reversible execution of the program. Automated optimization may target absolute minimization of memory or an optimal mix of memory and computation, achieving the best trade-off between memory save-restore operations and (re-)computation on a hardware platform-specific basis.

The **C** language is difficult to reverse as it is an advanced and complex language that incorporates decades of experience in efficient *forward-only* computing. Few efforts have been successful in developing a *correct* reversible execution of **C** programs in full generality. Plans for a reversible **C** execution were mentioned in past literature [Leeman, 1986] but no follow-up work on actual implementations could be found. The author of this book developed an initial version of a source-to-source compiler for a subset of the **C** language that successfully reversed certain classes of code fragments and applied the reversible code to efficient rollback in optimistic parallel discrete event simulation [Perumalla, 1999]. In another independent effort, the reversal of the **C**++ language was attempted but was not successful in producing correct results from reversible execution of large codes [Vulov et al., 2011]. In general, although there have been many attempts at *automated* program inversion, each with its own set of strengths and weaknesses, all reflect the principle of "no free lunch," either trading off generality [Glück and Kawabe, 2003, Kawabe and Glück, 2005] or efficiency [Dijkstra, 1979] or complexity or correctness [Eppstein, 1985, Srivastava et al., 2011a,b]. If a comprehensive solution is desired, automated reversal of any general-purpose high-level programming language remains as open research area as of this writing.

Admittedly, the challenge in achieving reversible execution arises from the necessity of revisiting every single language construct to examine how it affects reversibility, in isolation as well as in numerous potential combinations with other constructs. Thus, the key consideration to enabling reversibility would be *correctness* first, before *efficiency*, because there are many nuances in a programming language that, if not properly treated, can make the overall execution incorrect.

In the context of the **C** language, the classes of concepts to consider in combinations include the range of control flow constructs (the `,` (comma) operator, `if`, `while`, `do-while`, `for`, and `switch`); the pre-fix and post-fix operators inside expressions; and the variety of jump instructions (`goto`, `continue`, `break`, and `return`]. In arriving at a general-purpose, automated, and *correct* solution, it is impossible to discount any construct, and it is also not practical to rely on the programmer to provide the compiler-specific information, such as high-level algorithmic templates, for large and complex codes.

Interestingly, the parts of the language that deal with data types and their references are not affected in reversibility considerations, and they can be used unmodified. Thus, the aggregate data definitions using `struct` and `union` and naming via `typedef` can be retained unmodified from the original (irreversible) code, without any special transformations performed in the reversible code.

10.2   Source-to-Source Method for Reversible C

The compilation procedure follows the source-to-source methodology described in Section 9.3.1.1. The architecture of the reverse compilation is shown in Figure 10.1. In general, the application may be composed of portions that need to be reversible and the remainder of the application that uses the reversible portions. The portion of the application that needs to be made reversible is fed to the reverse compiler. For each original function given as input, the reverse compiler produces the code for the augmented forward version of the original function, along with its corresponding reverse function. The forward and reverse functions that are generated are combined with the application's customized runtime tape interface along with rest of the application that makes use of the forward and reverse functions. A conventional compiler is then invoked on all the pieces together to compile, link, and obtain the application executable.

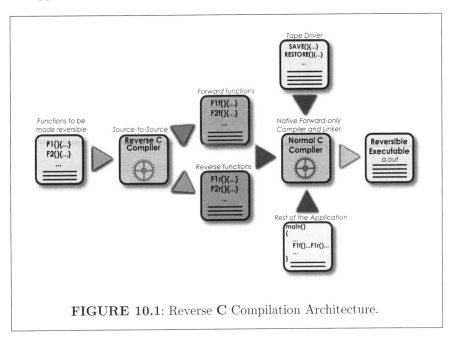

FIGURE 10.1: Reverse C Compilation Architecture.

10.2.1 Notation

In the code fragments to follow, the following notation is used.

For every function definition `F()` present in the source file input to the compiler, the compiler generates two versions of the function: `Ff()` and `Fr()`. The former is the original forward function suitably augmented for reversal, and the latter is the reverse version of the former. Also, the compiler assumes that this naming convention is obeyed by functions invoked from, but not defined in, the input files; the reverse code is generated based on this convention.

The term `s` represents any simple or compound statement, while `rs` denotes its reverse. In the transformed or generated code, extra braces are sometimes introduced to place certain statements, `s` or `rs`, within a new scope to (1) easily conform to the **C** rule that requires declarations to precede executable statements, and (2) prevent name conflicts among user-program variables and compiler-generated temporary variables.

In **C**, expressions can contain side-effect operations, such as `++` and `--`, embedded within them. Moreover, they have pre- or post-effect semantics, depending on context. For the purposes of source-to-source compilation, the side effects should be separated so that their relative orders are correctly reflected in the reversal. For any expression `test`, the prefix side-effect statements will be represented as `test-pre`, and the postfix side-effect statements as `test-post`, while the expression proper will be represented as `test-expr`.

10.2.2 Definition of Correctness of Reversible Execution

Reversibility of execution is achieved in the following sense of correct operation. For every function definition `F(){ ... }` given in the source files input to the compiler, the compiler generates `Ff(){ ... }` and `Fr(){ ... }` such that `{ Ff(); Fr(); }` is equivalent to a no-op. In other words, for every original function `F()`, the compiler generates forward function `Ff()` and reverse function `Fr()` such that the reverse function can be used to cancel all modifications performed by the forward function on the program's variables in memory.

In fact, a stricter sense of correctness is possible. The forward and reverse functions need not be invoked back-to-back for mutual cancellation, but may be intercepted by any additional intervening statements `s`, provided that `s` is reversed using `rs`. Thus, more generally, `{Ff(); s; rs; Fr();}` is a no-op.

The source-to-source method here assumes that if a reverse function is invoked, its forward function must have been invoked prior to the reverse. Commutativity is not guaranteed for the forward and reverse. In other words, `{ Fr(); Ff(); }` is not necessarily a no-op; in fact, in most cases, the results may be undefined because `Ff()` is necessarily one-to-one but not necessarily onto (i.e., injective but not necessarily bijective).

In regard to the execution model, a single thread of control flow is assumed in the program, with the traditional subroutine view: a stack of function calls formed by functions invoking other functions where the caller is suspended until the callee returns. Reversible execution of co-routines or multi-threading is not handled.

10.2.3   Runtime Tape Interface

A bit/byte tape data structure that follows last-in-first-out (LIFO) discipline is assumed to be part of the reversible runtime framework. A default implementation of this interface is provided by the compiler, but it can be customized or entirely replaced on an application-specific basis.

At least two tapes are assumed: one to record a trace of bits, and the other to record a trace of bytes. An implementation may choose to combine the two, but it may be more performance effective to keep the distinction between the bit tape and the byte tape. Both tapes must support two basic operations: SAVE and RESTORE.

A SAVE(x) operation results in appending the value of the *lvalue* x to the end of the tape, and RESTORE(x) retrieves the current value from the end of the tape and stores it into the *lvalue* x. The number of bits stored or retrieved is equal to 8*sizeof(x). A SAVE_BITS(x,n) operation results in saving only the lowest n bits of the integral *lvalue* x. Similarly, RESTORE_BITS(x,n) results in restoring only n bits into the integral *lvalue* x.

The bit tape is primarily used for conditional statements (to store the truth value) and for destructive assignments to bit variables, while the byte tape is used for other statements.

10.2.4   Compilation Phases

The compilation is divided into three phases, as illustrated in Figure 10.2. The phases delineate three distinct functionalities of the compiler. The *normalization* phase is needed to reduce the internal complexity of the compiler by mapping multiple, logically equivalent constructs into one canonical construct. The *transformation* phase is the process of actual generation of the forward and reverse code. The *optimization* phase encompasses the set of optimizations incrementally added to the compiler to increase the efficiency of reversal. Each of these phases is described in the following sections.

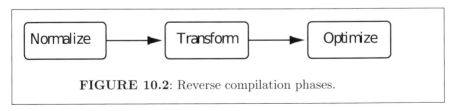

FIGURE 10.2: Reverse compilation phases.

10.3 Normalization

This phase, called the *normalization* process, is performed as the first step in the compilation process. To deal with the large set of language constructs, it is easier to first translate the program into a semantically equivalent program that uses a reduced set of language constructs. For example, all jump instructions (`case`, `break`, `return`) are translated to equivalent forms of the `goto` statement. Also, all iteration statements can be translated into equivalent forms of the `while` statement. Such a translation not only makes it easier to deal with a reduced set of instructions for the purposes of generating reverse code, but also makes optimizations on the smaller set easily applicable to all statements.

10.3.1 Declarations

Type declarations as well as global and static variable declarations are retained unchanged, as they are initialized at the beginning of program execution. Local variable declarations are also retained unchanged, except for moving their initializer expressions into explicit assignment statements, as illustrated in the following:

Original *Example*	Normalized *Example*
```\n{\n double x = 10, y = log(x);\n ...\n}\n```	```\n{\n double x, y;\n {\n x = 10;\n y = log(x);\n }\n ...\n}\n```

Initializer expressions in variable declarations are moved out of the declarations into actual assignment statements in the front of the code in which the variables are defined. Separation of initialization expressions from declarations makes it possible to retain the declaration statement proper unchanged in both forward and reverse code, and perform transformations and optimizations on initialization expressions automatically, as is done to the rest of the assignment expressions in the code outside declarations.

10.3.2 Side-Effect Expressions

Side-effect expressions are those that modify the variables while the expression is evaluated. The simplest example is the ++ operator that increments a variable as a side effect of accessing the variable's value. For example, the code `int x=10, y=x++;` results in y equal to 10, but x equal to 11 because x is incremented as a side effect of the expression `x++`. Operations with potential side effects include increment and decrement (pre-fix and post-fix), and comma operators. Calls to non-void functions inside expressions are also treated as side-effect expressions, unless the function call is the only term in the expression. All expressions that have side-effect operators are converted to equivalent set of expressions which do not have side-effects. In the preceding example, the side effect-free conversion is given by `int x=10, y=x; x++;` Expressions such as conditions, function call arguments, array indices, pointer arithmetic, etc., must be converted to be free of side effects. The elimination of side effects in expressions may add new temporary variables to hold intermediate values to make the expressions side effect-free. After the conversion, every side-effect operator appears as the only operator in a separate statement. This conversion makes it possible to decompose every expression into a deterministic sequence of updates to the variables.

10.3.3 Function Calls

All function calls are normalized such that a call to a `void` function appears as a statement by itself, and a non-`void` function call appears as a single right-hand-side term of an assignment statement:

$$\text{lvalue} = \text{function}(\cdots);$$

10.3.4 Arithmetic Expressions

To prevent loss of numerical precision during computation, all arithmetic expressions are normalized such that each value or variable is promoted to a data type of the next higher precision. This, of course, assumes that a special data type is available to the compiler that is of higher precision than the highest precision type used in the input functions. Otherwise, software emulation of higher precision is necessary, which can add a significant runtime overhead. Programs such as scientific codes predominantly containing floating point arithmetic require special treatment beyond this simple reversal approach. Advanced techniques will need to be employed, making the reversal less transparent (see Chapter 14).

10.3.5 `for` Statements

The `for` statement is normalized into its basic `while` form, as follows. Because the test predicate may contain pre-fix and post-fix operations embedded in the expression, these must be moved out of the expression. They are copied into the appropriate places to ensure that the semantics of the loop are maintained. All `continue` statements inside s are converted into `goto continue_label`. Similarly, `break` statements inside s are converted into `goto break_label`. Because s can be a non-compound statement, it is necessary to house it in a compound statement along with `incr` expression. A missing `test` expression is substituted with 1.

Original	Normalized

`for( init; test; incr )`
    `s`
	`{`
	   `init;`
	   `test-pre`
	   `while( test-expr )`
	   `{`
	     `test-post`
	     `s`
	     `continue_label:`
	     `incr;`
	     `test-pre`
	   `}`
	   `test-post`
	   `break_label:`
	`}`

Original *Example*	Normalized *Example*

`for( p=q; *(++q)!=*(p++); n++ )`
    `*p=*q;`
	`{`
	   `p=q;`
	   `++q;`
	   `while(*q!=*p)`
	   `{`
	     `p++;`
	     `*p=*q;`
	     `continue_label:`
	     `n++;`
	     `++q;`
	   `}`
	   `p++;`
	   `break_label:`
	`}`

10.3.6 `do-while` Statements

Each `do-while` loop is normalized in a manner similar to that for the `for` statement, as illustrated in the following.

Original	Normalized

`do`
  `s`
`while( test );`
	`{`
	  `s`
	  `test-pre`
	  `while( test-expr )`
	  `{`
	    `test-post`
	    `s`
	    `continue_label:`
	    `test-pre`
	  `}`
	  `test-post`
	  `break_label:`
	`}`

The treatment of `continue` and `break` statements inside s is the same as in the normalization of the `for` statement.

Original *Example*	Normalized *Example*

`do`
`{`
  `*p=*q;`
`} while( *(p++) == *(++q) );`
	`{`
	  `*p=*q;`
	  `++q;`
	  `while(*p==*q)`
	  `{`
	    `p++;`
	    `*p=*q;`
	    `continue_label:`
	    `++q;`
	  `}`
	  `p++;`
	  `break_label:`
	`}`

10.3.7 while Statements

Each `while` loop is normalized in a manner similar to that for the `for` statement, as illustrated in the following:

Original	Normalized

`while( test )`
  `s`
	`{`
	  `test-pre`
	  `while( test-expr )`
	  `{`
	    `test-post`
	    `s`
	    `continue_label:`
	    `test-pre`
	  `}`
	  `test-post`
	  `break_label:`
	`}`

The treatment of `continue` and `break` statements inside s is the same as in the normalization of the `for` statement.

Original *Example*	Normalized *Example*

`while( *(p++) == *(++q) )`
  `*p=*q;`
	`{`
	  `++q;`
	  `while(*p==*q)`
	  `{`
	    `p++;`
	    `*p=*q;`
	    `continue_label:`
	    `++q;`
	  `}`
	  `p++;`
	  `break_label:`
	`}`

10.3.8   return Statements

Multiple `return` statements are converted into a single `return` statement using the following normalization, which uses a temporary local variable for holding the returned value:

Original	Normalized

`T foo(...)`
`{`
  `...`
  `return x;`
  `...`
  `return y;`
  `...`
`}`
	`T foo(...)`
	`{`
	  `T t;`
	  `...`
	  `{t = x; goto label;}`
	  `...`
	  `{t = y; goto label;}`
	  `...`
	  `label:`
	  `return t;`
	`}`

In the case of a void function, the temporary variable can be dropped
After the normalization, at most one **return** statement will remain in any
function. This ensures a single exit point for each function, implying a single entry point for its reverse. In the transformation and reversal phases, no
further treatment is needed for the **return** statement. Hence it is retained
unchanged in the transformed code, and eliminated in the reverse code. Note
that the **label** name must be generated to be unique in the function scope.

10.3.9   continue **Statements**

The following normalization is performed on a **continue** statement by trans-
lating it into its equivalent **goto** statement.

Original	Normalized
`{`	`{`
`   . . .`	`   . . .`
`   continue;`	`   goto continue_label;`
`   . . .`	`   . . .`
`}`	`continue_label:`
	`}`

In the case of **while**, **for** and **do-while** statements, the position of
continue_label is already correctly placed during the normalization phase
to preserve the semantics of the **continue** statement.

10.3.10   break **Statements**

The following normalization is performed on **break** statements that are
present inside the bodies of **for**, **do-while**, and **while** statements. Each **break**
statement is normalized into an equivalent **goto** statement:

Original	Normalized
`{`	`{`
`   . . .`	`   . . .`
`   broak;`	`   goto break_label;`
`   . . .`	`   . . .`
`}`	`}`
`   . . .`	`break_label:`
`s`	`. . .`

Note that this normalization is *not* applied to the **break** statements that
belong to a **switch** statement. The **break** statements inside a switch statement
are handled differently, as described in Section 10.3.11 and Section 10.4.7.

10.3.11   switch **Statements**

The **switch** statement can be treated in two different ways for the purposes
of reversibility. In the first approach, it can be normalized into an equivalent
set of **if** and **goto** statements, as illustrated in the following:

Original	Normalized
| ```
switch( expr )
{
  case c1:
    s1
    break;
  case c2:
    s2
    /*fall-thru*/
  case c3:
    s3
    break;
  default:
    s4
    break;
}
``` | ```
{
  int c = expr;
  if( c == c1 ) goto label1;
  if( c == c2 ) goto label2;
  if( c == c3 ) goto label3;
  goto labeld;
label1:
  s1
  goto end;
label2:
  s2
  /*fall-thru*/
label3:
  s3
  goto end;
labeld:
  s4
  goto end;
end:
}
``` |

This normalization is a straightforward application of the semantics of a `switch` statement, that states that every case value acts as a label that becomes a target of a *computed goto* statement. Note that saving and restoring operations for the value of `c` will be introduced in the transformation phase of compound statements, as described in Section 10.4.4.

In the alternative approach, a `switch` statement can be transformed into an equivalent, but reversible, `switch` statement, as discussed later.

10.3.12 Post-Normalization State

At the end of the normalization, the following hold true:

- At most one assignment operator remains in any given expression statement. Moreover, both the left-hand-side and right-hand-side of every assignment operator are free of side effects.

- All `for` and `do-while` statements are eliminated by translating them into equivalent `while` statements.

- The condition expressions of `if`, `switch`, and `while` statements are free of side effects.

- The only type of jump statement remaining is the `goto` statement (with the exception of a single `return` statement at the end of the function body).

- There is at most one `return` statement in any function, which will be at the end of the function body. Moreover, in the case of non-void functions, the return expression consists of exactly one variable.

■ The relationship between unusual jumps across block scopes and the local variables in those blocks can be ignored in the transformation, because they are taken care of in the normalization. Thus, only normal entry into (and exit out of) block scopes need be considered.

10.4   Transformation

In the transformation phase, the input code is made reversible, and its reverse is also generated. The rules used in the transformation phase are discussed next. These rules are applied starting with each function definition in the input, and every statement in the function is recursively transformed.

10.4.1   Expression Statements

In **C**, the *constructive* operators include the following:

Constructive Operator	Inverse Operator	Description	Special Cases
++ (prefix)	-- (postfix)	Increment	Overflow bit
++ (postfix)	-- (prefix)	Increment	Overflow bit
-- (prefix)	++ (postfix)	Decrement	Underflow bit
-- (postfix)	++ (prefix)	Decrement	Underflow bit
+=	-=	Add to	Overflow bit
-=	+=	Remove from	Underflow bit
*=	/=	Multiply by	Zero operand
/=	*=	Divide by	Zero operand and non zero remainder
~=	~=	Bitwise not	None
^=	^=	Bitwise exclusive or	None

For all the constructive operators, no data needs to be saved to the tape by default (except for integer division, which requires the remainder to be saved). However, to take into account special situations such as overflow or underflow conditions zero operands, one bit must be logged on the tape for each constructive operator flagged with special cases. When the bit is zero, it indicates that no reversibility hazard is present, and hence nothing further is logged on the tape. If the bit is one, it indicates additional data must be obtained from the tape to properly reverse. In the case of overflow or underflow for increment or decrement operators, no additional data is needed, as the bit itself indicates sufficient information to recover from the operation properly.

All other operators are destructive assignments, which require the saving of the value of the operand before it is destroyed due to the operation. These operators include =, %=, and so on. Because all such operators are normalized into the form *lvalue op rvalue*, they are transformed to save the *lvalue* before the assignment:

Normalized	Transformed	Reversed

`lvalue op rvalue;`	`{`
	`    SAVE(lvalue);`
	`    lvalue op rvalue;`
	`}`
		`{`
		`    RESTORE(lvalue);`
		`}`

10.4.2   Function Calls

Because all function calls are normalized such that they are one of the forms `function(`···`)` or *lvalue*`= function(`···`)`, they are reversed by replacing the function name with the name of its reverse function.

To support function calls that are in fact evaluated on function pointer expressions, a hash table of function pointers is maintained at runtime by the compiler, and is used to map a function pointer to its reverse function pointer. Thus, if the function call is not via a simple function name, but is in fact via an expression that evaluates to a pointer to a function, then the hash table of function pointers is consulted to call the correct reverse function.

Normalized	Transformed	Reversed
`(func-ptr-expr)();`	`(func-ptr-expr)();`	`(lookup(func-ptr-expr))();`

10.4.3   Jump Statements

For each label that is the target of one or more `goto` statements in the input, the following transformation is applied.

Normalized	Transformed	Reversed
`{`	`{`	`{`
`...`	`char c;`	`char c;`
`s1`	`...`	`...`
`goto label;`	`s1`	`rs4`
`...`	`{c=1; SAVE(c);}`	`RESTORE(c);`
`s2`	`goto label;`	`switch( c )`
`goto label;`	`...`	`{`
`...`	`s2`	`case 0: break;`
`s3`	`{c=2; SAVE(c);}`	`case 1: goto rlabel1;`
`...`	`goto label;`	`case 2: goto rlabel2;`
`label:`	`...`	`}`
`s4`	`s3`	`...`
`...`	`...`	`rs3`
`}`	`{c=0; SAVE(c);}`	`...`
	`label:`	`rlabel2:`
	`s4`	`rs2`
	`...`	`...`
	`}`	`rlabel1:`
		`rs1`
		`...`
		`}`

This transformation works using a variable c to remember which of the goto statements was actually executed to reach the labeled statement in the forward execution. During reverse execution, this information is used to jump back to the point that immediately precedes the previously executed goto statement. To perform the reverse jump, the goto statements of the forward code are replaced by labels, which are reached using a *computed goto* in the reverse code.

This transformation works correctly even in the presence of multiple interdependent jump statements. It also correctly handles multiple visits to the same target label during the same function call. The SAVE() calls inserted before the goto statements are necessary to handle multiple visits to the same jump statement.

Note that the variable name c and the names of the labels in the reverse code—rlabel*i*—must be generated so that they are unique in the given function body scope. An easy way to do this is to use the label of the goto statement given in the original input. Also, note that the variables are declared to be of type char, assuming the number of goto statements for any label fits one byte (this is usually true for human-generated code). Otherwise, a larger integral type, such as short int, can be used.

Because the variable c is saved before the jump instruction, it need *not* be saved again when the variable goes out of scope, as is normally done for a compound statement as discussed next.

10.4.4 Compound Statements

Every compound statement (also called the block scope) is transformed according to the following illustration:

Normalized	Transformed	Reversed
| ```
{
    int x, y;
    s1
    s2
    s3
}
``` | ```
{
    int x, y;
    s1
    s2
    s3
    SAVE(x);
    SAVE(y);
}
``` | ```
{
    int x, y;
    RESTORE(y);
    RESTORE(x);
    rs3
    rs2
    rs1
}
``` |

By this transformation, the order of executable statements in the block is preserved in the forward code, but reversed in the reverse code. Because local variables lose their values when they go out of scope, they must be saved at the end of the block in the forward code, and restored at the beginning of the block in the reverse code. Hence, state saving calls are automatically added at the end of every compound statement body, so that the values of the local variables are saved for restoration during reverse execution. Abnormal exits from the block scope are treated specially (see Section 10.4.5).

Formal parameters to functions are logically similar to local variables of the function body. They are saved by treating them exactly like the local variables of the function body.

10.4.5 Jumps across Nested Blocks

In the case of unusual entry and exit via jump statements such as goto, the values of all the local variables that are about to go out of scope must be saved before the jump takes place. For this, the appropriate SAVE() calls are inserted before each goto, as illustrated in the following.

For example, a goto statement that takes the control out of a block scope must first save all the local variables in the block scope before transferring the control. Moreover, because it is possible that a single goto statement in a deeply nested block can cut across several block scopes, the local variables of all those blocks must also be saved before the jump is effected.

This is exactly analogous to the manner in which C^{++} generates calls to destructors on all local variables at all levels of block scope that are about to go out of scope as a result of any single goto statement [Ellis and Stroustrup, 1990].
```

Original	Normalized
`{/*C1*/`	`{/*C1*/`
`  int x;`	`  int x;`
`  ...`	`  ...`
`  {/*C2*/`	`  {/*C2*/`
`    int y;`	`    int y;`
`    ...`	`    ...`
`    {/*C3*/`	`    {/*C3*/`
`      int z;`	`      int z;`
`      ...`	`      ...`
`      goto label;`	`      {SAVE(z); SAVE(y);}`
`      ...`	`      goto label;`
`    }`	`      ...`
`  }`	`    }`
`  ...`	`  }`
`label:`	`  ...`
`  ...`	`label:`
`}`	`  ...`
	`}`

In the preceding illustration, the `goto` statement would result in the variables y and z going out of scope. Hence, their values are saved before the jump is executed. Note that the variable x is *not* saved because it remains in scope even after the jump, and hence will be saved by instructions generated by the normal transformation phase (see Section 10.4.4).

10.4.6 if Statements

Each `if` statement is made reversible by introducing a variable to remember the result of the `if` condition. The value of this variable is saved after the `if` statement is executed during the forward execution. The saved value is restored during reverse execution, and it is used to recollect which of the true or false branches was executed in the forward execution. The corresponding branch is then executed in reverse. The transformation is as follows:

Normalized	Transformed	Reversed
`if( test )`	`{`	`{`
`    s1`	`  char c = !!test;`	`  char c;`
`else`	`  if( c )`	`  RESTORE(c);`
`    s2`	`    s1`	`  if( c )`
	`  else`	`    rs1`
	`    s2`	`  else`
	`  SAVE(c);`	`    rs2`
	`}`	`}`

In practice, the `SAVE(c)` and `RESTORE(c)` are replaced by `SAVE_BITS(c,1)` and `RESTORE_BITS(c,1)`, as the actual value of c is only contained in the lowest bit. Note that the save and restore operations on c are actually generated as part of the transformation phase on compound statements, as described in Section 10.4.4. Note also that it is necessary to coerce the truth value of the condition to 0 or 1 to make it fit in one bit. The coercion is necessary because C permits any non-zero value to qualify as a true value, which obviously does not fit in one bit. Double negation (using the ! operator twice) is one way to perform the coercion, as illustrated in the transformed code.

The result of the if condition must be saved on the tape *after* its true or false branch is executed. This is because the number of bits required of either branch is generally unknown in advance, and hence makes it difficult during reverse execution to find out which branch to reverse. Saving the condition value after the branches are executed makes the condition value become the last item on the tape at the time of reversing the if statement, which can be readily popped from the tape.

An optimization can be performed by eliminating the added variable in case the if condition is invariant in both true and false branches of the if statement. This is because the same expression can be reevaluated (without loss of information) during reverse execution to recollect which branch was executed, as illustrated in the following:

Normalized	Transformed	Reversed

`if( condition )`
`    s1`
`else`
`    s2`
	`{`
	`    if( condition )`
	`        s1`
	`    else`
	`        s2`
	`}`
		`{`
		`    if( condition )`
		`        rs1`
		`    else`
		`        rs2`
		`}`

Even though only a single bit is reduced by such optimization per if statement execution, such optimization can result in a significant reduction in overall memory utilization when the statement is repeatedly executed, say, inside a loop.

10.4.7   switch Statements

One way to treat a switch statement is to normalize it into if and goto statements, as described in Section 10.3.11. An alternative technique is to transform it into another switch statement, which uses two integer variables, c and d, as illustrated in the following:

Original	Transformed	Reversed
| ```
switch( expr )
{
  case v1:
    s1
    break;
  case v2:
    s2
    /*fall-thru*/
  case v3:
    s3
    break;
  default:
    s4
    break;
}
``` | ```
{
  char c = 0, d = 0;
  switch( expr )
  {
    case v1: c=1; d++;
      s1
      break;
    case v2: c=2; d++;
      s2
      /*fall-thru*/
    case v3: c=3; d++;
      s3
      break;
    default: c=0; d++;
      s4
      break;
  }
  SAVE(c);
  SAVE(d);
}
``` | ```
{
  char c, d;
  RESTORE(d);
  RESTORE(c);
  switch( c )
  {
    case 0:
      rs4
      if(--d <= 0) break;
    case 3:
      rs3
      if(--d <= 0) break;
    case 2:
      rs2
      if(--d <= 0) break;
    case 1:
      rs1
  }
}
``` |

The variable c serves to normalize the case values to fall in the range $0..n-1$. Because the case expression is permitted to be of any integral type, the normalization serves to reduce the memory requirements to remember the switch value (assuming the expression is not invariant in the switch body). The variable d serves to record the number of case labels that are traversed during the forward execution. This is necessary because **C** permits fall-through execution in the absence of **break** statements between case labels, and because **break** statements could potentially be present inside conditional statements.

Note that the save and restore operations on c and d are actually generated as part of the transformation phase on compound statements, as described in Section 10.4.4.

The flag and fall-through counter variables used in the transformation must be saved on the tape *after* the entire body of the **switch** statement is executed. This enables the counter values to be recovered easily from the end of the tape before the reversal of the **switch** statement begins.

10.4.8 while Statements

Each **while** statement is made reversible using a counter variable to remember the actual number of iterations that were executed. This local variable is saved after the execution, and restored before reversal, as illustrated in the following:
```

Normalized	Transformed	Reversed

`while( condition )`
    `s`
	`{`
	    `int c = 0;`
	    `while( condition )`
	    `{`
	        `c++;`
	        `s`
	    `}`
	    `SAVE(c);`
	`}`
		`{`
		    `int c;`
		    `RESTORE(c);`
		    `while( c > 0 )`
		    `{`
		        `rs`
		        `--c;`
		    `}`
		`}`

It is important to increment the counter `c` *before* executing `s`, as it is possible for `s` to contain jump instructions that take control out of the `while` body. Note that this transformation makes the `while` statement reversible even in the presence of any jump instructions (`break`, `continue`, `goto` and `return`). This is as a result of combining the normalization of `break`, `continue` and `return` statements (see Section 10.3.10, Section 10.3.9, and Section 10.3.8, respectively) with the transformation of `goto` statements (see Section 10.4.5 and Section 10.4.4).

Note that the save and restore operations on `c` are actually generated as part of the transformation phase on compound statements, as described in Section 10.4.4.

Similar to `if` statements, the counter variables for the loop statements must be saved on the tape *after* the loop has ended. This ensures that the iteration count is readily available as the last item on the tape at the time of reversing the loop statement.

10.4.9 Libraries and I/O

The **C** standard library is an integral runtime component of the language. Due to the large number of routines in the standard library interface, the reversibility of the functions in standard libraries is not addressed here. However, it is possible to reverse the standard libraries by applying the compilation methodology to the source code of the library function implementation, and by following our naming scheme to map each standard library function to its reverse peer.

Reversal of input and output is possible by redefining the semantics of the file operations. In particular, swap semantics are needed to properly and comprehensively address the reversibility of input and output. This is a major undertaking because it touches on the I/O subsystem interfaces beyond the basic language semantics, and thus would require a more elaborate infrastructure. Moreover, transparent backward compatibility to traditional semantics of forward-only computing can be extremely difficult to achieve. While the high-level idea of reversible I/O is known, effective and efficient implementation in a complex environment such as **C** standard I/O library is the subject of ongoing for future research.

10.4.10   Pragmas

To reduce the effort needed in the development of the first version of the
compiler prototypes, some programmer involvement may be warranted. The
programmer may use a set of pragmas to annotate portions of the code to
aid reversibility. This helps relieve the burden of tedious (but mechanical)
compiler code development, at least until production versions of the source-
to-source compiler are developed that automatically glean such reversibility
information from the input code. Pragmas can either be guarantees of certain
conditions or hints to the compiler for optimization. For guarantees, pragmas
equivalent to syntactic reversibility of **Janus** could be employed. For example,
the post-branch condition analogous to that of a **Janus** conditional statement
(see Section 8.3.3.2) could be attached via a pragma string at the end of the
conditional statement in **C**.

The list of example pragmas is shown in Table 10.1. Some of these pragmas can be eliminated (or automatically inserted by a smart preprocessor)
using sophisticated compiler techniques. The pragmas are only necessary to
reduce the amount of effort in compiler implementation; a production version
of the compiler can eliminate the need for such pragmas. For example, the
INVARIANTCONDITION pragma for an if statement can be eliminated in a sophisticated compiler implementation that can perform advanced dependency
analysis between the if condition and the if branches.

10.5   Optimization

Several optimizations are possible after the input code is subjected to the
normalization and transformation phases of compilation. The higher the sophistication of the compiler with respect to the optimizations, the better the
efficiency of the forward and reverse code. Some optimizations are described
next.

10.5.1   Value Recovery or Reconstruction

Although a local view of the code may indicate that a value is being destroyed,
it is possible that a copy of the same value is either available in an unmodified form at another location, or the value can be reconstructed by some
recomputation.

10.5.2   Irreversible and Environment Slices

In general, certain operations performed in the input code must be ignored
for reversal purposes. There are two reasons for this:

TABLE 10.1: List of Example Pragma Specifications

Pragma	Description	Example
MAP_REVERSE=f,r	Specifies r as the reverse function of f. If r is specified as void, all calls to f in the forward code will be removed in the reverse code.	`#pragma MAP_REVERSE=SAVE,RESTORE` `#pragma MAP_REVERSE=printf,void`

INVARIANT_ CONDITION	Specifies that the condition of the immediately following if statement is invariant in both its true and false branches. This eliminates the need to save the truth value to the tape.	`#pragma INVARIANT_CONDITION`
		`if( ... )`
		`{`
		`    ...`
		`}`

RECOVERABLE= *var-list*	Specifies that the given local variables are recoverable, and hence do not need to be saved before going out of scope. This pragma must appear immediately before the compound statement begins.	`#pragma RECOVERABLE=w,z`
		`{`
		`    int w, x, y, z;`
		`    ...`
		`}`

BEGIN_GLOBAL	Marks pragmas appearing between BEGIN_GLOBAL and END_GLOBAL to be globally applied.	`#pragma BEGIN_GLOBAL`
		`  #pragma MAP_REVERSE=printf,void`
		`  #pragma MAP_REVERSE=max,max`
		`#pragma END_GLOBAL`

STATELESS_SLICE	Marks the immediately following compound statement to consist exclusively of stateless code. This implies saving to tape can be turned off for destructive assignments, and local variables need not be saved before they go out of scope. However, the reversal of statements and reversal of function calls are performed.	`#pragma STATELESS_SLICE`
		`{`
		`    ...`
		`}`

1. Irreversible operations: In any program, certain operations can be assumed to be irreversible for practical purposes. For example, output statements may not be reversed. A message-send over the network cannot always be physically withdrawn.

2. Environment: In certain applications, some operations are indeed reversible but the application has other (probably more efficient) means of reversing those operations. For example, in optimistic parallel simulation, the retraction of a events scheduled by a rolled back event is performed by the parallel simulation kernel. Such operations must be ignored during reverse compilation.

Without loss of generality, both of the preceding types of operations can be deemed to be irreversible for the purposes of reverse compilation.

The slices of the program that only lead to such irreversible operations can be *sliced away* during reversal. In other words, for those portions of the forward computation that solely lead to irreversible computation: (1) no attempt is made to make them reversible during forward computation, and (2) such code is ignored during reverse computation. Applying this optimization can result in reduced overheads for reversibility.

Example: Consider the following code fragment in which the value of a variable y is computed and printed:

Original	Unoptimized		Optimized	
	Transform	Reverse	Transform	Reverse
`{`	`{`	`{`	`{`	`{`
`  int x;`	`  int x;`	`  int x;`	`  int x;`	`  int x;`
`  s1`	`  s1`	`  RESTORE(x);`	`  s1`	`  RESTORE(x);`
`  {`	`  {`	`  rs4`	`  {`	`  rs4`
`    int y;`	`    int y;`	`  {`	`    int y;`	`  {`
`    s2`	`    s2`	`    int y;`	`    s2`	`    rs3`
`    y = f(x);`	`    y = f(x);`	`    RESTORE(y);`	`    y = f(x);`	`    rs2`
`    print(y);`	`    print(y);`	`    rs3`	`    print(y);`	`  }`
`    s3`	`    s3`	`    rs2`	`    s3`	`  rs1`
`  }`	`    SAVE(y);`	`  }`	`  }`	`}`
`  s4`	`  }`	`  rs1`	`  s4`	
`}`	`  s4`	`}`	`  SAVE(x);`	
	`  SAVE(x);`		`}`	
	`}`			

The value of y is computed as a function of x, and printed as output. The slice of the program that only leads to the irreversible printing consists of the variable y and the statements y=f(x); and print(y);. All the components of this slice can be safely eliminated in the reverse code.

10.5.3   Eliminating Reversal of Initialization

Even though initializations are destructive assignments, they never need to be undone, and hence do not need to be checkpointed. This is because initially the variables contain *don't care* values, whose exact values are immaterial for the correctness of the program. For reversal, the same variables can hold any values, which are not necessarily the same *don't care* values that were held before initialization, as any value trivially qualifies as a *don't care* value. For example, consider the following code fragment, and its unoptimized and optimized reverse codes:

Original	Unoptimized		Optimized	
	Transform	Reverse	Transform	Reverse
`{`	`{`	`{`	`{`	`{`
`  int x;`	`  int x;`	`  int x;`	`  int x;`	`  int x;`
`  x = 0;`	`  SAVE(x);`	`  RESTORE(x);`	`  x = 0;`	`  RESTORE(x);`
`  s`	`  x = 0;`	`  rs`	`  s`	`  rs`
`}`	`  s`	`  RESTORE(x);`	`  SAVE(x);`	`}`
	`  SAVE(x);`	`}`	`}`	
	`}`			

The unoptimized version saves the pre-initialization value of x before initializing it to zero, and restores x to the saved value during reverse execution toward the last statement. Because there is no other statement in the reverse code that depends on the pre-initialization value of x, such restoration is useless. Hence, both the saving and restoration code for pre-initialization values are eliminated in the optimized version.

This optimization can result in significant memory savings, as local variables are very commonly declared and initialized inside functions.

Note that this optimization is not restricted to local (stack) variables, but is also applicable to variables allocated on the heap. The same applies to memory buffers whenever they are re-initialized for irreversible reuse. In fact, a simple generalization of this optimization is: For memory items that contain *don't care* values, (destructive) assignment of values to them need not be reversed.

10.5.4   Invariant Expressions

Determining whether an expression is invariant in a block of code is useful in performing certain optimizations to reduce overheads. For example, no condition variable needs to be added for an `if` statement if it is true that the expression is invariant in both branches of the `if` statement (i.e., the condition expression is guaranteed to evaluate to exactly the same value both in forward and in reverse). This fact can be determined by checking the bodies of the true and false branches of the `if` statement to see if any of the variables used in the condition are indeed modifiable in the branches. If none of the variables contained in the condition are affected in the bodies, then the result of the condition need *not* be saved, but instead the exact same condition expression can be reevaluated to determine the condition value in both forward and reverse codes. In case the condition expression is not invariant in the body of the `if` statement, the result of the condition needs to be saved during the forward execution. However, for this purpose, only one bit is needed to remember the truth value of the condition. Transformations for both cases (variant and invariant) are described in Section 10.4.6.

Similarly, for a `switch` statement, an additional variable is introduced to remember the value of the `switch` expression, lest it is changed in the body of the `switch`. This additional variable can be eliminated if the `switch` expression can be determined to always be invariant in the `switch` body.

10.5.5   Common Sub-Expression Elimination

Although the reverse compiler increases the number of sub-expressions and variables in order to render all expressions side effect-free, the potential increase in sub-expressions is mitigated by the fact that common sub-expression elimination will later be performed by the usual **C** compiler when the generated code is compiled, and no saving is performed for initialization of the new variables.

10.5.6   Switch Statement Trade-Offs

A switch statement can be either normalized into `if` and `goto` statements, or can be transformed into another `switch` statement. The choice between the two can be made at compile time, depending on which of the two results in a smaller memory requirement for saving.

10.5.7   Tape Compression

Run-length encoding and decoding can be performed on the tape. This can result in significant memory savings, for example, in case of loops involving conditional statements. Consider the following code pattern that commonly occurs in programs:

```
done = 0;
while( !done )
{
  if( expr )
  {
    done = 1;
  }
  else
  {
    ...
  }
}
```

In the preceding code fragment, the loop is executed until a condition `expr` becomes true. If the condition expression is modified in the `if` branches, then its truth value (1 bit) must be saved for each iteration (see Section 10.5.4).

Suppose the loop is executed n times. It is clear that all the first $n-1$ bits will hold a 0 value, and only the last bit will contain a value 1. If run length encoding is used on the tape, then the first $n-1$ bits can be reduced to $O(\log_2 n)$ bits.

10.6   Tape Size Upper Bounds

A simple set of translation rules that can be used by the compiler is shown in Table 10.2. The most common types of statements used in high-level languages are listed, along with their corresponding instrumented and reverse code outputs. Against each of the statements, the state size that is achievable for that statement type is listed. Because not all operations of the input model are perfectly reversible, it is necessary to add control state information to be able to reverse them. However, the better the understanding of the semantics of the code, the better the ability to reduce the state size. Hence, the reduction in state size can vary, depending on the sophistication of the compiler. The translation rules of Table 10.2 thus place an *upper bound* on the state size, which could potentially be improved via optimization.

The instrumented forward computation code, as well as reverse code, are generated by recursively applying the rules of Table 10.2 to the input model. The significant parts of these rules are their state bit size requirements and the reuse of the state bits for mutually exclusive code segments. Each of the rules is described in detail next.

- **T0**: The if statement can be reversed by taking note of which branch is executed in the forward computation. This is done using a single bit variable **b**, which is set to 1 or 0 depending on whether the predicate evaluated to true or false in the forward computation. The reverse code can then use the value of **b** to decide whether to reverse the if part or the else part when trying to reverse the if statement.

  Because the bodies of the if part and the else part are executed mutually exclusively, the state bits used for one part can also be used for the other part. Hence, the state bit size required for the if statement is one plus the larger of the state bit sizes, x_1, of the if part and x_2 of the else part, that is, $1 + max(x_1, x_2)$.

- **T1**: Similar to the simple if statement (**T0**), an n-way if statement can be handled using a variable **b** of size $\log_2(n)$ bits. Thus, the state size of the entire if statement is $\log_2(n)$ for **b**, plus the largest of the state bit sizes, $x_1 \ldots x_n$, of the component bodies, that is, $\log_2(n) + max(x_1 \ldots x_n)$ (as the component bodies are mutually exclusive).

- **T2**: Consider an n iteration loop, such as a for statement, whose body requres x state bits for reversibility. Then n instances of the x bits can be used to keep track of the n instances of invocations of the body, giving a total of $n * x$ bit requirement for the loop statement. The inverse of the body is invoked n times in order to reverse the loop statement.

- **T3**: A loop with variable number of iterations, such as a `while` statement, can be treated the same as a fixed iteration loop, but the actual number of iterations executed can be noted at runtime in a variable b. The state bits for the body can be allocated based on an upper limit n on the number of iterations. Thus, the total state size added for this statement is $\log_2(n) + n * x$.

- **T4**: For a function call, no instrumentation is added. For reversing it, its inverse is invoked. The inverse is easily generated using the rules for **T7** described later. The state bit size, x, is the same as for **T7**.

  In the simple case in which the function call graph is a tree, the state bit sizes can be completely determined *statically*. In the case of models in which the function call graph is a directed acyclic graph (DAG), the (maximum) state bit size requirements can still be statically determined. However, in the more general case of an arbitrary function call graph (implying the presence of direct and/or indirect recursion), it is difficult to statically determine the maximum state bit sizes.

- **T5**: Constructive assignments, such as `++`, `--`, `+=` and so on, do not need any instrumentation in normal cases. The reverse code uses the inverse operator, such as `--`, `++`, `-=`, respectively. Most of these constructive statements do not require any state for reversibility, but some need one bit to note overflow or underflow conditions.

- **T6**: Each destructive assignment, such as `=`, `%=`, and so on, can be instrumented to save a copy of its left-hand side into a variable b before the assignment takes place. The size of b is $8k$ bits for assignment to a k-byte left-hand side variable (*lvalue*).

- **T7**: In a sequence of statements, each statement is instrumented depending on its type, using the previous rules. For the reverse code, the sequence is reversed, and each statement is replaced by its inverse, again using the corresponding generation rules from the preceding list. The state bit size for the entire sequence is the sum of the bit sizes of each statement in the sequence.

- **T8**: Determining the state size requirements in the presence of arbitrary jump instructions is difficult, because in general it could be unbounded. In simple cases such as one in which the no `goto` label in the model is reached more than once during an event computation, the following analysis can be used. Because at most one index per `goto` label is stored, the bit size requirement of this scheme is $\log_2(n + 1)$, where n is the number of `goto` statements that are the sources of that single target label. Note that even if a label is the target of only one jump instruction, at least one bit is required to distinguish between reaching the label normally (falling-through) and reaching the label as a result of the jump instruction.

■ **T9**: Any legal nesting of the previous types of statements can be treated by recursively applying the corresponding generation rules. The state bit size is also obtained by the corresponding state-bit composition rule.

10.7 Tape Size Determination

Table 10.2 shows the generation rules and upper bounds on state size requirements for supporting reverse computation. s, or $s_1..s_n$ are any of the statements of types T0..T7. rs is the reverse code of the statement s. b is the corresponding saved bits "belonging" to the given statement. The operator `=@` is the inverse operator of a constructive operator `@=`, (e.g., `-=` for `+=`).

To determine the amount of state needed to reverse a given computation, the following procedure is used. Because the input code is a sequence of statements, start with T7 (or, alternatively, T4), and recursively apply the rules of Table 10.2. This is done while reusing the bits on code segments that are mutually exclusive (as indicated by the MAX() operation in the table).

It can be observed from the table that the statements with potentially higher state bit sizes are destructive assignments, nestings of conditional statements within loops, nested loops inside loops, and destructive operations among interdependent jump instructions.

TABLE 10.2: Summary of State Bit Sizes for Various Statement Types

Type	Description	Application Code			Bit Requirements		
		Original	Instrumented	Reverse	Self	Child	Total
T0	simple choice	if() s_1; else s_2;	if() {s_1; b=1;} else {s_2; b=0;}	if(b==1) {rs_1;} else {rs_2;}	1	x_1, x_2	$1 + max(x_1, x_2)$
T1	compound choice	if() s_1; elsif() s_2; elsif() s_3; else() s_n;	if() {s_1; b=1;} elsif() {s_2; b=2;} elsif() {s_3; b=3;} else {s_n; b=n;}	if(b==1) {rs_1;} elsif(b==2) {rs_2;} elsif(b==3) {rs_3;} else {rs_n;}	$\log_2(n)$	$x_1, x_2, ..., x_n$	$\log_2(n) + max(x_1, ...)$
T2	fixed iterations (n)	for(n) s;	for(n) s;	for(n) rs;	0	x	nx
T3	variable iterations (maximum n)	while() s;	b=0; while() {s; b++;}	for(b) rs;	$\log_2(n)$	x	$\log_2(n) + nx$
T4	function call	$f()$;	$ff()$;	$fr()$;	0	x	x
T5	constructive assignment	v @= w;	v @= w;	v =@ w;	1	0	1
T6	k-byte destructive assignment	v = w;	{b = v; v = w;}	v = b;	$8k$	0	$8k$
T7	sequence	s_1; s_2; s_n;	s_1; s_2; s_n;	rs_n; rs_2; rs_1;	0	$x_1 + ... + x_n$	$x_1 + ... + x_n$
T8	jump (label lbl as target of n goto's)	goto lbl; s_1; goto lbl; s_n; lbl: s;	b=1; goto lbl; s_1; b=n; goto lbl; s_n; b=0; lbl: s;	rs; switch(b) { case 1: goto $label_1$; case n: goto $label_n$; } rs_n; $label_n$; rs_1; $label_1$:	$\log_2(n+1)$	0	$\log_2(n+1)$
T9	Nestings of T0-T8	Apply the above recursively			Apply the above recursively		

Chapter 11

Reversal of Linear Codes

11.1   Automated Generation

The method of determining reverse code based on local information, such as inverting each statement individually, is effective in many situations, but it is not the most efficient (memory-wise or computationally). In contrast, other non-local approaches are needed to arrive at a good reversal, ideally, eliminating the need for history. Examples include backtracking in the *def-use graph* of the variables and/or performing inter-statement code analysis. How much higher one might have to rise in increasing the scope or reach of code depends on the particular code to be reversed. In this vein, reversibility depends on the beholder's eye, especially on the granularity of computation at which the reversal is attempted. While individual instructions might not be retraced backward perfectly, some program fragments might be perfectly invertible when considered in aggregate. In other words, while a local view might be irreversible, a more global view might readily make it reversible with reduced or no memory. However, the generation of the aggregate reverses may be highly challenging.

An excellent illustration of this local versus non-local reversal approach is with a class of functions called *linear codes*, which are sequences of assignments of arbitrary linear expressions to variables. It turns out that even the simplified problem of reversing an assignment-sequence (linear control flow) requires a non-trivial algorithm. However, it is indeed possible to reverse linear codes despite the presence of destructive assignments and apparent singularities in the input code.

11.2   Example: Fibonacci Sequence

Consider the functions in Algorithm 11.1 used to generate Fibonacci sequences of the form $X_n = X_{n-1}+X_{n-2}$, where, say, $X_0 = 1$ and $X_1 = 2$. The algorithm defines a function init() that initializes the sequence, with a= X_0 and b= X_1. A function fib() advances the sequence one step forward. The sequence can be generated by invoking fib() multiple times. After k invocations of fib(), the variable a contains X_{k+1} and b contains X_{k+2}. The challenge we consider here is the automatic generation of the inverse revfib() that uncomputes fib(). In other words, any k invocations of fib() followed by the same number of invocations of revfib() restores a and b to the same values as they had before the first invocation of fib(). This would provide the capability to move forward as well as backward in the sequence. In general, we would like an automated method, preferably one that can be incorporated into a compiler, to automatically generate inverse functions for such forward functions.

Algorithm 11.1 Functions for reversible Fibonacci sequence generation

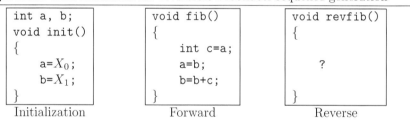

| Initialization | Forward | Reverse |

When we attempt to generate the reverse code line-by-line, we quickly encounter two instances of the *destructive assignment* problem: the assignments c=a and a=b, each of which needs history to save the old values to reverse. In general, a major difficulty arises from destructive assignments because they are not invertible when considered in isolation. A destructive assignment is one in which a variable is overwritten with a new value that has no correlation with the variable's old value when that assignment statement is examined in isolation from its preceding code.

However, the Fibonacci sequence is indeed mathematically reversible. Hence, it should be possible to generate a reverse function without requiring history despite the presence of destructive assignments. Here, we will illustrate how a method based on a expanded global view can relieve the limitation of a local view.

11.3 General Linear Codes

Consider the class of codes called *linear codes*, of which the Fibonacci sequence generator example is but a special case. This class encompasses several other common operations such as swap, circular/destructive shift, and circular/lossy rotation. Linear codes are a generalization of basic operators that are reversible, and include many additional operators. The swap operation, which is a self-reverse, is in fact a special case with $n = 2$ of a circular rotation of n variables. The circular rotation itself is a *linear code* that can be represented by an invertible matrix of constants operating on the vector of variables [Perumalla, 2003].

Let $L(\ldots)$ denote any linear expression of variables and constants. For any variable v, let v' denote the value of that variable just prior to the first modification in forward execution.

A linear code is a sequence $S = (s_1, \ldots, s_m)$ of assignments $s_i : v_i \leftarrow L_i(1, v_1, \ldots, v_n, v'_1, \ldots, v'_n)$ of linear expressions L_i on any variables. The linear expression can span both old values as well as new values of the variables to allow for multiple updates to the same variable, and it can also use constants in the expression. Variables can be assigned multiple times or not at all. Local variables can also be used. However, jump instructions, branch statements, loops, or recursion are not allowed. The function must consist of only a single sequence of assignment statements.

11.4 Linear Code Reversal Algorithm

We will now present an algorithm for reversal, that takes any linear code and generates its reverse code. The main idea behind the algorithm is the observation that the linear sequence of assignments can be represented as a matrix product operation. Indeed, let the old values of the variables be represented as a column vector V' and the operations of the linear code assignments be represented as a matrix of constants W. Then the matrix product WV' gives a new column vector corresponding to the new values of the variables $V \leftarrow WV'$. This indicates that we only need to multiply both sides by the inverse of W. The resulting equation $V' \leftarrow W^{-1}V$ recovers the old values in terms of the current values. Thus, all we would need to do for reversal is to apply the inverse of W. The meat of the algorithm is then concerned with addressing singularities when W as obtained from the user-written code happens to be non-invertible. The steps to deal with all the cases are listed in Algorithm 11.2.

For an input linear code, the algorithm generates equivalent forward and reverse codes such that the reverse function exactly restores the values of variables changed by the forward function. This is achieved with memory space whose size is independent of the number of invocations (either consecutive or intermixed) of the forward and reverse functions.

Algorithm 11.2 Linear code reversal algorithm

1. Preprocess forward code

2. Obtain matrix representation

3. Iteratively eliminate matrix singularity:

    (a) Perform row elimination
    (b) Perform column elimination

4. Invert matrix

5. Generate optimized reverse code

11.4.1 Preprocessing

The first step is to convert the arbitrary user-specified sequence of assignments into a different, equivalent version in which there is at most one assignment per variable. This is also called Static Single Assignment (SSA) in the compiler literature. The preprocessing is shown in Algorithm 11.3.

11.4.2 Matrix Representation

After preprocessing, let $L_i = w_{i0} + \sum_j^n w_{ij} v_j'$, $1 \le i \le n$. Then, S can be written as $V \leftarrow WV'$ corresponding to

$$
\begin{bmatrix} 1 \\ v_1 \\ \vdots \\ v_n \end{bmatrix} \leftarrow \begin{bmatrix} 1 & 0 & \cdots & 0 \\ w_{10} & w_{11} & \cdots & w_{1n} \\ \vdots & \vdots & \cdots & \vdots \\ w_{n0} & w_{n1} & \cdots & w_{nn} \end{bmatrix} \begin{bmatrix} 1 \\ v_1' \\ \vdots \\ v_n' \end{bmatrix}.
$$

If W is non-singular, then it is easy to recover V' by multiplying both sides of the equation by the inverse of W: $V' \leftarrow W^{-1}V$.

For the Fibonacci sequence generator `fib()` of Algorithm 11.1, the series of transformations performed by the algorithm is shown in Figure 11.1. The resulting forward and reverse functions are listed in Algorithm 11.4.

Algorithm 11.3 Preprocessing in linear code reversal algorithm

- **Input** User-specified sequence of m^0 assignment statements

$$S^0 \equiv (s_1^0, \ldots, s_{m^0}^0)$$

on n^0 variables, where

$$s_i^0 \equiv v_i \leftarrow L_i^0(1, v_1, \ldots, v_{n^0}, v_1', \ldots, v_{n^0}').$$

Note that any v_i could appear as the left-hand side (LHS) of zero or more assignment statements, and hence the m^0 can be larger or smaller than n^0.

- **Goal** Convert the input function to an equivalent function, with possibly fewer or more assignment statements, such that

  1. Each variable appears as the LHS of exactly one assignment.
  2. All right-hand side (RHS) expressions are rewritten equivalently in terms of values held by each variable immediately prior to the first assignment statement in the function.

- **Algorithm**

  1. Temporarily treat local variables as global.
  2. For every variable v_i, add the assignment $v_i \leftarrow v_i'$ to the top of the function, that is, $S^0 \leftarrow (v_i \leftarrow v_i', S^0)$.
  3. For every assignment s_i^0, replace it with

$$v_i \leftarrow L_i^0(1, E(v_1), \ldots, E(v_{n^0})),$$

   where $E(v)$ is the RHS expression of the most recent assignment to v. In other words, in every linear expression L, expand every reference to v by its equivalent expression that is only based on old values v' and not on any new values.

  4. At this point, no global variable will have dependency on any local variable. For every local variable v_l, delete all assignments to v_l.
  5. From S^0, for every assignment $v_i \leftarrow \ldots$, delete all except the last assignment to v_i.

- **Output** A revised sequence of assignments $S = (s_1, \ldots, s_m)$, where $s_i : v_i \leftarrow L_i(1, v_1, \ldots, v_{i-1})$ of linear expressions L_i only on variables already assigned values, and no v_i appears more than once on the LHS of any assignment.

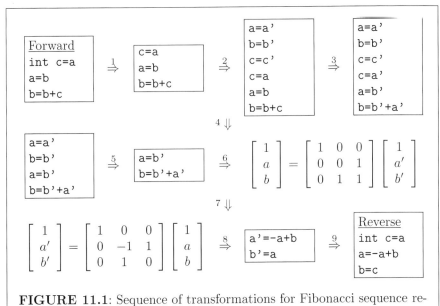

FIGURE 11.1: Sequence of transformations for Fibonacci sequence reversal.

Algorithm 11.4 Automatically generated reversal for Fibonacci sequence

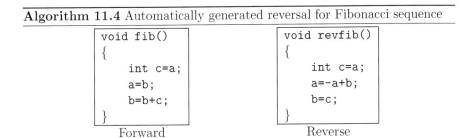

Forward	Reverse

11.4.3 Eliminating Singularity

When the matrix W is invertible, as in the Fibonacci example, it is straightforward to recover the old values V' of the variables V. But what if W turns out to be singular, with no inverse? This case is addressed next.

When W is singular, at least one of the rows can be expressed as a linear combination of the other rows, or at least one of the columns can be expressed as a linear combination of the other columns. Consider the simple linear code in Algorithm 11.5 on two variables a and b.

Clearly, the matrix is non-invertible. However, this does *not* imply that the code cannot be reversed. In fact, one of the possible reverse codes is
a'=a/3; b'=2a'; .

In general, the solution to the singularity problem is to eliminate all lin-

---

Algorithm 11.5 Example code containing singularity preventing reversal

Forward:	```a=a+b;``` ```b=2a;```
Preprocessed:	```a=a'+b';``` ```b=2a'+2b';```
Matrix:	$\begin{bmatrix} 1 & 0 & 0 \\ 0 & 1 & 1 \\ 0 & 2 & 2 \end{bmatrix} \begin{bmatrix} 1 \\ a' \\ b' \end{bmatrix} = \begin{bmatrix} 1 \\ a \\ b \end{bmatrix}$

---

ear combination relationships among the variables or sub-expressions. In the preceding example, the RHS expressions of both the variables a and b are simply linearly related to each other, which makes the resultant matrix non-invertible. Removal of the linear dependencies can be accomplished via either row elimination or column elimination. Between the two choices, row elimination may be preferable because it does not add any new variables, thereby minimizing memory usage.

Let W be represented as a column matrix of rows R_i, $0 \leq i \leq n$, as

$$W = \begin{bmatrix} R_0 \\ \vdots \\ R_n \end{bmatrix}.$$

When W is singular, one of the variables, say v, can be expressed as a linear combination of the other variables. Using this fact, we can eliminate v from the RHS of all other variables. This reduces the matrix size from $n \times n$ to $(n-1) \times (n-1)$. This process can be repeated until the matrix is not singular. In the preceding singularity example, the second row can be expressed as two times the first row:

```
a=a'+b';
b=2a;
``` $\Rightarrow$ ```b'=2a';``` $\Rightarrow$ ```a=a'+2a';``` $\Rightarrow$ ```a=3a';``` $\Rightarrow$ ```a'=a/3; b'=2a';``` .

An analogous process can be performed to eliminate columns instead of rows. While row elimination corresponds to the capture of the redundant variable, column elimination essentially corresponds to the capture of the redundant common sub-expression that occurs on the RHS expressions of all the variables. Using the column elimination approach, we can observe that the sub-expression a'+b' occurs on the RHS of both variables. This can be corrected by introducing a new variable, c=a+b, and transforming the forward code as

```
a=a'+b';
b=2(a'+b');
``` $\Rightarrow$ ```a=a'+b'; b=2(a'+b'); c=a+b;``` $\Rightarrow$ ```a=c'; b=2c'; c=3c';``` $\Rightarrow$ ```a'=c''; b=2c''; c'=c/3; c''=c'/3;``` $\Rightarrow$ ```c'=c/3; a'=c/9; b'=2c/9;```

Note the use of c'' in the derivation. Although we can compute c', we still need to recover the user's original values of a' and b'. Because c'=a'+b' gives only one equation with two unknowns, we cannot recover a' and b' from c' alone. Computing c'' and reevaluating a' and b' from c'' solves the problem.

Thus, there are two equivalent reversals possible for the same forward code.

Guards are needed to deal with boundary conditions for the initial values. These guard conditions protect the reverse execution from going backward past the valid forward-reverse code relationship. This can require protecting from 0 to n initial conditions, where n is the number of variables in the forward code.

11.5   Fast Backward

A natural extension of the matrix approach is in reversing by more than one step backward at a time. While normal reverse execution mode is treated for a single step backward, it is possible to efficiently jump s steps backwards at once. Because the update proceeds as $V \leftarrow WV'$, the state after s steps is given by $V \leftarrow W^s V'_s$, where V and V'_s are values of the variables s updates apart. The fast reversal backward by s steps is achieved as $V'_s \leftarrow (W^s)^{-1}V$.

For example, the Fibonacci sequence can be jumped two steps at a time with $s = 2$ using the following derivation:

$$
\begin{bmatrix} 1 \\ a \\ b \end{bmatrix} = \begin{bmatrix} 1 & 0 & 0 \\ 0 & 0 & 1 \\ 0 & 1 & 1 \end{bmatrix} \begin{bmatrix} 1 \\ a' \\ b' \end{bmatrix}
$$

$$
\Rightarrow \begin{bmatrix} 1 \\ a \\ b \end{bmatrix} = \begin{bmatrix} 1 & 0 & 0 \\ 0 & 0 & 1 \\ 0 & 1 & 1 \end{bmatrix} \begin{bmatrix} 1 & 0 & 0 \\ 0 & 0 & 1 \\ 0 & 1 & 1 \end{bmatrix} \begin{bmatrix} 1 \\ a'' \\ b'' \end{bmatrix}
$$

$$
\Rightarrow \begin{bmatrix} 1 \\ a \\ b \end{bmatrix} = \begin{bmatrix} 1 & 0 & 0 \\ 0 & 1 & 1 \\ 0 & 1 & 2 \end{bmatrix} \begin{bmatrix} 1 \\ a'' \\ b'' \end{bmatrix}
$$

$$
\Rightarrow \begin{bmatrix} 1 & 0 & 0 \\ 0 & 2 & -1 \\ 0 & -1 & 1 \end{bmatrix} \begin{bmatrix} 1 \\ a \\ b \end{bmatrix} = \begin{bmatrix} 1 \\ a'' \\ b'' \end{bmatrix}
$$

$$
\Rightarrow \boxed{\texttt{a''=2a-b; b''=-a+b;}}
$$

The reverse code to jump backward two steps at a time is given in Algorithm 11.6.

The computation of the inverse matrix $(W^s)^{-1}$ can be time-consuming when the number of variables is large. However, the number of variables is not large for programs whose source codes are human generated. Moreover, the determination of the inverse matrix is a one-time cost, incurred at compilation time, not runtime.

Algorithm 11.6 Single- and double-stepping reverse functions for Fibonacci sequence

void fib()
{
    int c=a;
    a=b;
    b=b+c;
}
1-step Forward
	void revfib()
	{
	    int c=a;
	    a=-a+b;
	    b=c;
	}
	1-step Reverse
		void revfib2()
		{
		    int c=a;
		    a=2a-b;
		    b=-c+b;
		}
		2-step Reverse

11.6 Other Common Linear Codes

■ **Swap**: The swap operation $x \leftrightarrow y$ is a simple, small case of linear codes. When written in matrix form, it gives a matrix that is a self-inverse.

$$
\boxed{
\begin{array}{l}
\text{Forward} \\
\text{int t=x;} \\
\text{x=y;} \\
\text{y=t;}
\end{array}}
\Rightarrow
\boxed{
\begin{array}{l}
\text{t=x';} \\
\text{x=y';} \\
\text{y=x';}
\end{array}}
\Rightarrow
\boxed{
\begin{array}{l}
\text{x=y';} \\
\text{y=x';}
\end{array}}
\Rightarrow
\begin{bmatrix} 1 \\ x \\ y \end{bmatrix}
=
\begin{bmatrix} 1 & 0 & 0 \\ 0 & 0 & 1 \\ 0 & 1 & 0 \end{bmatrix}
\begin{bmatrix} 1 \\ x' \\ y' \end{bmatrix}
$$

$$
\Rightarrow
\begin{bmatrix} 1 & 0 & 0 \\ 0 & 0 & 1 \\ 0 & 1 & 0 \end{bmatrix}
\begin{bmatrix} 1 \\ x \\ y \end{bmatrix}
=
\begin{bmatrix} 1 \\ x' \\ y' \end{bmatrix}
\Rightarrow
\boxed{
\begin{array}{l}
\text{x'=y;} \\
\text{y'=x;}
\end{array}}
\Rightarrow
\boxed{
\begin{array}{l}
\text{Reverse} \\
\text{int t=x;} \\
\text{x=y;} \\
\text{y=t;}
\end{array}}
$$

■ **Circular Rotate**: A circular rotation of an array of variables is a linear code. Each element is shifted as $x_i \leftarrow x'_{(i+b \bmod n)}$, $0 \leq i < n$, where n is the number of elements, $b = 1$ for left shift, $b = -1$ for right shift, x'_i is the pre-shift value and x_i is the post-shift value. This is a generalization of the swap operation ($n = 2$ gives the swap). For example, a left shift would result in

$$
\begin{bmatrix} 1 \\ x_0 \\ x_1 \\ \vdots \\ x_{n-2} \\ x_{n-1} \end{bmatrix}
=
\begin{bmatrix}
1 & 0 & 0 & 0 & \cdots & 0 & 0 \\
0 & 0 & 1 & 0 & \cdots & 0 & 0 \\
0 & 0 & 0 & 1 & \cdots & 0 & 0 \\
\vdots & \vdots & \vdots & \vdots & \ddots & \vdots & \vdots \\
0 & 0 & 0 & 0 & \cdots & 0 & 1 \\
0 & 1 & 0 & 0 & \cdots & 0 & 0
\end{bmatrix}
\begin{bmatrix} 1 \\ x'_0 \\ x'_1 \\ \vdots \\ x'_{n-2} \\ x'_{n-1} \end{bmatrix},
$$

which is easily inverted as

$$
\begin{bmatrix}
1 \\
x_0' \\
x_1' \\
\vdots \\
x_{n-2}' \\
x_{n-1}'
\end{bmatrix}
=
\begin{bmatrix}
1 & 0 & 0 & 0 & \cdots & 0 & 0 \\
0 & 0 & 0 & 0 & \cdots & 0 & 1 \\
0 & 1 & 0 & 0 & \cdots & 0 & 0 \\
\vdots & \vdots & \vdots & \vdots & \ddots & \vdots & \vdots \\
0 & 0 & 0 & 0 & \cdots & 0 & 0 \\
0 & 0 & 0 & 0 & \cdots & 1 & 0
\end{bmatrix}
\begin{bmatrix}
1 \\
x_0 \\
x_1 \\
\vdots \\
x_{n-2} \\
x_{n-1}
\end{bmatrix}.
$$

Chapter 12

Reversible Random Number Generation

12.1 Random Stream Traversal: Forward versus Reverse

Probability distributions are routinely employed in computer codes such as simulations of physical system models. Random number streams are used to generate samples that conform to certain probability distributions. In computationally intensive simulations, very large numbers of random samples are drawn (millions to billions in number).

The generation of, or traversal along, pseudorandom number streams or sequences has been very well studied over many years. Traditionally, aspects of considerable interest in random number generation have been (1) maximizing or increasing the lengths of random number streams for a given precision ot each number, (2) increasing the computational speed of generation, (3) approximation techniques for complex distributions, and (4) reducing correlations across multiple parallel streams. However, the aspect of *reversibility* of generation is a less-studied consideration. While it intersects orthogonally to most of these factors, it has not yet received as much attention. Thus, the majority of literature in random number generation is predominantly concerned with the *forward* mode of traversal along a random number stream, but much less so with the *reverse* mode.

In the context of reversible computing, if a forward computation invokes a random number generator (RNG), and that computation is later reversed, it is necessary to reverse the random number generation as well. Otherwise, results from subsequent forward execution become non-deterministic and unrepeatable; they also deviate from a corresponding irreversible execution, making it harder to validate and accept the final answers from the computation.

In order to be able to correctly roll back the random number generation, traditionally, the seed value is checkpointed after every generated random number sample. This does solve the correctness problem but introduces a large amount of inefficiency because of the memory copying time and the consumption of a large amount of memory. The alternative is the reverse computation approach in which the inverse code from the forward code of the RNG can be developed and invoked to go backward in the stream. The challenge, then, is to determine the inverse code of the forward RNG code.

To avoid checkpointing, if the reverse compilation approach is attempted, the following problem is quickly encountered. RNGs rely on lossy/destructive assignments such as modulo operations. This implies that a straightforward application of automation techniques can degenerate to (incremental) checkpointing, and they are not any more effective than checkpointing approaches. To get around this problem, manually developed reversible RNGs are needed that do not require any checkpointing to backstep along the random stream. On an abstract level, such RNGs can indeed be expected to exist because, after all, random number streams are statically laid out cyclic sequences of numbers, and it should be possible to traverse forward and backward along the cycles with the same ease. Fortunately, many RNGs do possess forward and reverse generation functions (e.g., [L'Ecuyer and Andres, 1997]) that can be easily applied for reverse computation.

The need for avoiding checkpointing and focusing on reverse computation approaches to reversible random number generation is more pronounced when one considers some of the recent high-quality RNGs. Due to new statistical requirements arising out of their large size and inherent complexity, recent simulation applications require RNGs with stronger statistical properties and longer periods, which together imply an increase in the seed size of the RNG. For example, the "Mersenne Twister" (MT19937) RNG has an extremely long period of $2^{19937} - 1$, and is of high statistical quality [Matsumoto and Nishimura, 1998]. However, it requires 624 words of space for seeds. Whenever random number generation is performed in the forward execution, the seeds must be saved. For a traditional forward computation using this generator, 2496 bytes of state would need to be saved per forward invocation just to support the undo operation for the RNG. One might think that *incremental* checkpointing could be employed here, but the way this RNG is structured, portions of bits from each word are subject to change every time a random number is generated, thus making it difficult to optimize using incremental checkpointing techniques. Assuming the reverse recurrence can be found for MT19937, which its creators believe is possible, no checkpointing is required for the seeds, which reduces the memory requirements of such applications considerably. Because of the reduction in checkpointing overheads, system performance will improve as well.

Here we will examine some efficient reversible methods that permit both the traditional *forward* traversal as well as *backward* traversal. Reversible random number generators are presented for four common classes of probability distributions, as summarized in Table 12.1.

12.2 Memory-Based Method to Make a Generator Reversible

For simple distributions, the random number generation procedure is often specified by a closed-form formula (e.g., for the exponential distribution). More complex distributions are either computationally complex or may not be expressible as closed-form expressions. While the simpler ones may be reversed more easily than the others, there is one general method that is applicable to any generator.

Reversal of any random number generator can be accomplished by simply storing the generated bits in computer memory (persistent or volatile storage, depending on desired type of usage), and replaying them for reversal. Thus, the simplest reversal is obtained by logging all forward sequences of random numbers, and reading them from the log in last-in-first-out (LIFO) order during reversal. This is the most memory intensive, but the most general, method because it is not restricted to computationally generated random numbers. For example, random numbers sourced from atmospheric radiation can be easily reversed using this memory log-based approach. Thus, in the case of hard-to-reverse physically based random number generators or other hard-to-invert pseudorandom generators, reversibility can be accomplished simply by recording the conventional forward sequence in memory. The memory can be filled in advance of use, or can be generated on-the-fly during use. Either way, the amount of memory available to store the generated numbers determines the length of the random sequence that can be reversed (i.e., traversed in either direction). Given a memory size limit of M_z bits and a precision of b bits per

TABLE 12.1: Summary of Reversible Random Number Generators

Forward	Reverse	Memory -Limited	Description
$\mathcal{R}_{\mathcal{T}}()$	$\mathcal{R}_{\mathcal{T}}^{-1}()$	Yes	Memory-based generator that can wrap any underlying forward generator $\mathcal{R}^*()$ and make it reversible, within memory limits
$\mathcal{R}_{\mathcal{U}}()$	$\mathcal{R}_{\mathcal{U}}^{-1}()$	No	Generator for the uniform distribution
$\mathcal{R}_{\mathcal{C}}()$	$\mathcal{R}_{\mathcal{C}}^{-1}()$	No	Generator for distribution with cumulative distribution function (CDF) $c(x) = \int_0^x p(\overline{x})d\overline{x}$ whose inverse $c^{-1}(\cdot)$ is analytically known, where $c^{-1}(c(x)) = x$
$\mathcal{R}_{\mathcal{J}}()$	$\mathcal{R}_{\mathcal{J}}^{-1}()$	No	Acceptance/Rejection-based generator for any distribution with known probability density function (PDF) $p(\cdot)$ and "cover" function $u(x)$ such that $p(x) \leq u(x)$ for all x, and the cover function has an invertible CDF so that it can be randomly sampled using a suitable generator of type $\mathcal{R}_{\mathcal{C}}()$
$\mathcal{R}_{\mathcal{K}}()$	$\mathcal{R}_{\mathcal{K}}^{-1}()$	No	Upper- and Lower-bound-based Acceptance/Rejection generator for any distribution with known PDF $p(\cdot)$ and a series of K upper-bound functions $u_k(\cdot)$ and lower-bound functions $l_k(\cdot)$ such that, for $0 \leq k < K$, $p(x) \leq u_{k+1}(x) \leq u_k(x)$ and $l_k(x) \leq l_{k+1}(x) \leq p(x)$, and $u_0(x)$ can be sampled using a suitable generator of type $\mathcal{R}_{\mathcal{C}}()$

number, the length of the sequence that can be reversed at any given moment is $M = M_z/b$.

Algorithm 12.1 shows a scheme to traverse forward and backward along any random number sequence. The conventional forward random sequence is generated by calls to a function $\mathcal{R}^*()$. The inverse of $\mathcal{R}^*()$ is assumed to be either unknown, or physically infeasible to obtain, or expensive to compute. The algorithm uses a circular buffer B of size M to keep a reversal window of M numbers, counting backward from the latest call to $\mathcal{R}^*()$. The circular buffer is realized as an array indexed from 0 to $M-1$, with indices h, c and t used to remember the head, current, and tail positions, respectively, in the reversal window. The user of the random number stream must invoke $\mathcal{R}_\mathcal{T}()$ instead of $\mathcal{R}^*()$ to generate the next random number in the forward sequence. To step backward in the sequence, the user invokes $\mathcal{R}_\mathcal{T}^{-1}()$. Any order of the forward and backward calls can be made, as long as at most M consecutive calls are made to $\mathcal{R}_\mathcal{T}^{-1}()$ after the most recent call to $\mathcal{R}^*()$. After $n \geq 0$ calls to $\mathcal{R}^*()$, the reversal window is of length $min(n, M)$.

Algorithm 12.1 Reversing any forward random generator $\mathcal{R}^*()$ using a finite M-sized circular buffer

---

$B[]$=Array indexed 0 to $M-1$, used as circular buffer of M b-bit integers
M=Buffer size ($M \geq 1$)
h=Leading position in (index into) circular buffer, initially 0
c=Current position in (index into) circular buffer, initially 0
t=Lagging position in (index into) circular buffer, initially $M-1$

$\mathcal{R}_\mathcal{T}()$:	$\mathcal{R}_\mathcal{T}^{-1}()$:
if $h = c$ **then**	$c \leftarrow (c-1) \mod M$
$\quad B[h] \leftarrow \mathcal{R}^*()$	$u \leftarrow B[c]$
$\quad h \leftarrow (h+1) \mod M$	**return** u
$\quad$ **if** $h = t$ **then**	
$\quad\quad t \leftarrow (t+1) \mod M$	
$\quad$ **end if**	
end if	
$u \leftarrow B[c]$	
$c \leftarrow (c+1) \mod M$	
return u	
(a) Forward	(b) Reverse

This scheme can be used for any generator (either a pseudorandom number generator or a physically based generator), at the expense of storage space. Also, the reversal window is necessarily limited by the amount of memory devoted to this generator. Note that the size (b bits) for each element in the buffer corresponds to the size of the variate and not the size of the underlying generator state. For example, in the case of pseudorandom number generators, the bit size b_s of the underlying seed can be much larger than b bits required to represent one double precision number. In such cases, the size of the trace obtained by saving the double precision number is much smaller than a trace in which the sequence of seeds is saved. For very high-quality random numbers, b_s can be as large as 1024 bits or even larger, whereas typically $b \leq 64$. In certain generators that generate bits rather than numbers, $b = 1$, M may have to be set to a large value because such generators typically generate several millions of random bits per second.

12.3 Pseudorandom Numbers

In the absence of a universally accepted definition of randomness, any reference to randomness should, in a strict sense, be interpreted as pseudorandomness. Nevertheless, the term "pseudorandom numbers" is commonly used to specifically refer to deterministic cycles of integers generated from within a closed computing system. They are designed for ease of use, and for repeatable reproduction of sequences. In practice, the bit precision b of the generated integers is sufficiently large ($b \geq 32$), and the cycle lengths P are also large (e.g., $P \geq 2^{128}$). Because the generators are essentially cycles, it is conceptually possible to traverse the cycles forward as well as backward, although often generators are only written for forward traversal of the cycles.

12.3.1 Forward Generation

The operation of the generators is envisioned as follows. The user invokes $\mathcal{R}^*()$ to generate the next number in the random number sequence. The internal state, such as the memory used by the variables, of the generator is encapsulated as s. For example, the set of seeds and any other working variables together constitute the state of the generator. Without loss of generality, the memory size for s can be considered to remain constant throughout the random number sequence. A function $g(s)$ maps the state to a number r in a range desired by the user for random numbers. Note that the bit width of r is bounded by (and often much less than) the bit width of s. The state s is then overwritten by the value returned by a function $f(s)$ that generates the random sequence of evolution.

The literature is rich with a wide range of pseudorandom number generators, each defining a pair of $f(\cdot)$ and $g(\cdot)$ functions (see, for example, [Gentle, 2003]). In most generators, $f(\cdot)$ is a computationally complex sequence of arithmetic operations, while $g(\cdot)$ is a relatively simpler function. However, other generators also exist in which the opposite is true, in which $f(\cdot)$ can be as simple as an increment operation, while $g(\cdot)$ assumes the entire functionality of randomization.

12.3.2 Reversible Generation

For reversal of the random number sequence, the user invokes a new function $\mathcal{R}^{*-1}()$ that is defined to take one step backward in the sequence and recover the previous number. In general, $n \geq k$ consecutive invocations of $\mathcal{R}^*()$, for any $k \geq 0$, followed by $n - k$ consecutive invocations of $\mathcal{R}^{*-1}()$, has the aggregate effect on the stream that is the same as the effect of k consecutive invocations of $\mathcal{R}^*()$ alone.

Let $s \leftarrow f^{-1}(s)$ be the inverse operation of the update $s \leftarrow f(s)$ to recover the previous state of the seed s. If $f^{-1}(\cdot)$ is known, then it is easy to invert the generator in $\mathcal{R}^*()$ and arrive at a procedure for $\mathcal{R}^{*-1}()$. However, often it is challenging to find $f^{-1}(\cdot)$ for any given $f(\cdot)$. In general, there is no known method to create a reversible version of a conventional forward pseudorandom number generator.

The property of pseudorandom number codes that makes it challenging to discover their inverses automatically is that the generators contain destructive assignments, thereby *apparently* making them irreversible. Yet, it is conceptually reasonable to expect that they ought to be perfectly reversible with zero memory trace because the cycles ought to be amenable to traversal in the opposite direction to that of the forward generation. One straightforward way to achieve reversal is by relying on the knowledge of the period of the sequence (i.e., after how many invocations a seed value repeats itself in the sequence). Then, the reverse can be produced by fast-forwarding and looping back in sequence, rather than by true reversal of execution path. For example, if the period (length of the cycle) is P, then executing $s \leftarrow f(s)$ exactly $P - 1$ times ensures that s moves one step backward. Under modulo arithmetic, it is possible to realize the repeated application of the arithmetical operations efficiently.

Another way to define reversible random number generators is by restricting the operations of $f(\cdot)$ to only perfectly reversible versions (e.g., using only constructive operations such as accumulation and lossless bit shift operators). However, we are not aware of generators that are defined using only constructive operations. Cellular automata-based generators [Gentle, 2003] appear possible to adopt for this purpose, based on the constructive aspect of updates based on neighbor values, but reversibility aspects of boundary conditions need to be accounted for in complete reversibility.

12.4   Reversible Generation from the Uniform Distribution

One of the most primitive random streams is one that produces variates uniformly distributed in the interval $[0, 1]$. A large number of other complex distributions can be built over random samples from this uniform distribution. For this reason, the uniform random number generator is an essential building block for most other schemes. Thus, in order to make other complex distributions reversible, it is first necessary to develop reversible generators for the uniform distribution. A uniform distribution is given by a probability distribution function $p(\cdot)$ in the interval $[a, b)$ satisfying the following equation for its cumulative distribution function (CDF) $c_p(\cdot)$:

$$c_p(x) = \int_a^x p(\overline{x})d\overline{x} = \begin{cases} \left(\frac{x-a}{b-a}\right) & \text{if } a \leq x < b \\ 1 & \text{if } x \geq b \end{cases}.$$

The template for forward and reverse execution of a uniform pseudorandom number generator is shown in Algorithm 12.2. The function $\mathcal{S}()$ computes the next seed values from the current seed values, while $\mathcal{S}^{-1}()$ computes the previous seed values from the current seed values. The function $\mathcal{U}()$ maps the seed values to the range $[0, 1)$. Note that the same function $\mathcal{U}()$ is used in both forward and reverse algorithms.

Algorithm 12.2 Uniform random number generator

$\mathcal{R_U}()$:	$\mathcal{R_U}^{-1}()$:
$u \leftarrow \mathcal{U}(s)$ $s \leftarrow \mathcal{S}(s)$ **return** u	$s \leftarrow \mathcal{S}^{-1}(s)$ $u \leftarrow \mathcal{U}(s)$ **return** u
(a) Forward	(b) Reverse

12.4.1   Open versus Closed Ranges

To facilitate reversibility, the uniform random number must be defined to be sampled from $[0, 1)$, That is, all real numbers between 0 (inclusive) and 1 (exclusive). If it is instead generated in $[0, 1]$, that is, with both 0 and 1 being inclusive, it usually interferes with one-to-one properties required for reversal.

For example, consider an operation of the form

$$\theta \leftarrow (\theta + r \cdot 2\pi) \mod 2\pi$$

in which an angle $\theta \in [0, 2\pi)$ is rotated by a random offset $r \cdot 2\pi$, modulo 2π, where r is a variate from the uniform distribution in the interval between 0 and 1. This operation can be reversed as

$$\theta \leftarrow (\theta - r \cdot 2\pi) \mod 2\pi.$$

However, if $r \in [0, 1]$ (that is, inclusive on both ends), then both $r = 0$ and $r = 1$ result in the same value for θ, violating the one-to-one mapping requirement for reversal, and resulting in loss of information and ambiguity for reversal. In effect, the range $[0, 1]$ maps 0 and 1 to the same value, which eventually creates ambiguity. Thus, it is best to restrict the uniform distribution to the interval $[0, 1)$ rather than to the $(0, 1)$ that is exclusive on both ends or $[0, 1]$ that is inclusive on both ends. The use of $(0, 1)$ can also be envisioned without interfering with reversibility, but the exclusion of 0 is not usually desired.

12.4.2   Linear Congruential Generators

A popular class of pseudorandom number generators is the Linear Congruential Generator (LCG). In an LCG, $f(\cdot)$ is a function of the form

$$x_{i+1} = (ax_i + c) \mod m,$$

where a and m are integers, $0 < a < m$ and $0 \le c < m$, and c and m are mutually prime. The term a is a primitive root of m, that is, for every integer $0 < d < m$, there exists an integer k such that $a^k \mod m = d$.

An LCG can be reversed using the inverse of a (modulo m) given by b as

$$b = a^{m-2} \mod m,$$

giving

$$x_i = (bx_{i+1} - c) \mod m.$$

In all the preceding equations, the modulus operation is defined on any integer $m > 0$ and (positive or negative) integer x as

$$x \mod m = \begin{cases} x & \text{if } 0 \le x < m, \\ (x - m) \mod m & \text{if } m \le x, \text{ and} \\ (x + m) \mod m & \text{if } x < 0. \end{cases}$$

The forward and reverse functions, $\mathcal{S}$ and $\mathcal{S}^{-1}$, respectively, for linear congruential generators for the uniform distribution are given in Algorithm 12.3.

For example, consider $m = 7$ whose primitive root is $a = 3$. The inverse of a is $b = 3^{7-2} \mod 7 = 5$. Let $c = 2$. For these LCG parameters, the sequence in Table 12.2 is obtained using the forward and reverse procedures of Algorithm 12.3.

Algorithm 12.3 Reversible linear congruential generator

Variables	Forward	Reverse
x {Seed} m {Modulus} a {Multiplier} c {Increment} $b \leftarrow a^{m-2} \mod m$	$\mathcal{S}()$: $x \leftarrow (ax+c) \mod m$	$\mathcal{S}^{-1}()$: $x \leftarrow (b(x-c)) \mod m$

TABLE 12.2: Example LCG Sequence for $m = 7$, $a = 3$, and $c = 2$

i	x_i	Forward $x_{i+1} \leftarrow (ax_i + c) \mod m$	Reverse $x_i \leftarrow (b(x_{i+1} - c)) \mod m$
0	x_0	↓ 5	↑ 5
1	x_1	↓ 3	↑ 3
2	x_2	↓ 4	↑ 4
3	x_3	↓ 0	↑ 0
4	x_4	↓ 2	↑ 2
5	x_5	↓ 1	↑ 1
6	x_6	↓ 5	↑ 5

12.4.3 Counting-Based Generators

A special class of random number generators exists in which $f(\cdot)$ is a computationally trivial function, making it easy to arrive at their inverses. In such cases, it is straightforward to define the forward-reverse pairs of functions to move forward and backward in the random number stream. For example, with the *counter-based* generators [Salmon et al., 2011], this is achieved by simply incrementing or decrementing the seed value, respectively, to move forward and backward, respectively, in the stream. In addition to sequential traversal of the random sequence, it is also possible with such counter-based methods to directly jump to any arbitrary place in the sequence. Thus, reversal of such sequences becomes a trivial operation, namely using a single counter variable to remember the current location, and incrementing or decrementing it as needed to move forward or backward in the sequence. This simple reversal algorithm for counting-based generators is illustrated in Algorithm 12.4.

Algorithm 12.4 Update function $\mathcal{S}()$ for counting-based generators

Forward	Reverse
$\mathcal{S}(s)$: $s \leftarrow s + 1$	$\mathcal{S}^{-1}(s)$: $s \leftarrow s - 1$

12.5 Reversible Generation from Invertible Cumulative Distributions

In the case of simple probability distributions such as the exponential or Pareto distributions, closed-form inversions of their CDF are known. Such distributions can be easily traversed in forward and reverse directions by a simple procedure that is built using reversible uniform random generators. Reversing the sampling operation on an exponential distribution thus becomes as simple as invoking the reversal of the underlying uniform random number generator, once per reversal step. The restoration of the uniform random number seed is necessary and sufficient for reversing the sampling of the distribution.

The typical method based on invertible CDF works as follows. Given the CDF $c_p(\cdot)$ of a PDF $p(\cdot)$, the distribution can be sampled as $x_r = c_p^{-1}(r)$, where r is uniformly distributed in $[0, 1]$. Because this gives a direct mapping from the uniform distribution to the desired distribution, the reversible generation of x_r is a simple mapping from the reversible generation of r. For example, if the PDF is the exponential distribution

$$p(x) = \begin{cases} \lambda e^{-\lambda x} & \text{if } x \geq 0, \\ 0 & \text{otherwise,} \end{cases}$$

then the CDF is given by

$$c_p(x) = \begin{cases} 1 - e^{-\lambda x} & \text{if } x \geq 0, \\ 0 & \text{otherwise,} \end{cases}$$

and the inverse CDF is given by

$$c_p^{-1}(r) = x_r = \begin{cases} \frac{-\log(1-r)}{\lambda} = \frac{-\log r}{\lambda} & \text{if } r \in [0, 1], \\ 0 & \text{otherwise.} \end{cases}$$

The forward and reverse algorithms of random number generation for any distribution with an inverse CDF $c_p^{-1}(r)$ are shown in Algorithm 12.5.

12.6 Reversible Generation from Probability Density Functions

There are many instances in which, for a given PDF, the integral for its CDF cannot be analytically determined. In other cases, the CDF is computationally expensive to evaluate for every sample. In such cases, algorithmically complex sampling procedures are used, involving control flow such as conditional statements and iteration. The reversal challenge for such distributions is rooted in

Algorithm 12.5 Random number generator for distributions with invertible cumulative distribution functions

$\mathcal{R}_C()$:	$\mathcal{R}_C{}^{-1}()$:
$r \leftarrow \mathcal{R}_\mathcal{U}()$ $x_r \leftarrow c_p^{-1}(r)$ **return** x_r	$r \leftarrow \mathcal{R}_\mathcal{U}{}^{-1}()$ $x_r \leftarrow c_p^{-1}(r)$ **return** x_r
(a) Forward	(b) Reverse

the fact that the control flow breaks the one-to-one correspondence between the underlying uniform random number seed stream and the probability distribution sample stream.

12.6.1   Reversibility Problem

Here we consider a class of approaches called *acceptance-rejection* methods often used to generate samples from any complex distribution that is specified only by its PDF. In the context of reversibility, the most important aspect of these methods is that they take multiple samples from the uniform distribution for generating each single sample of the complex distrbution. A variable number of calls to the uniform random number generator are made in a loop to examine a sequence of candidates before selecting one that satisfies the distribution of interest.

For reversal to properly step back to a previous sample, it is necessary to execute the iteration in reverse order as many times as was performed in the forward execution. For example, if the forward execution took i_n iterations of the loop to generate the nth sample, the loop must be reversed the same i_n number of times in the reverse execution. Within each iteration, among other things, one sample (or, in general, a fixed number of samples) of the uniform distribution is taken. This results in a logical mapping from the one sample of the complex distribution to one or more samples of the uniform distribution. This dynamically changing nature of mapping from the nth sample of the complex distribution to the number of samples i_n of the uniform distribution introduces a challenge for reversible execution. The variable mapping is illustrated in Figure 12.1, in which each vertical line represents a sample taken from the distribution by the generator.

If the general automation methods described in Chapter 9 are applied to the code of the forward generator for the complex distribution, the automated method will log the number of iterations i_n to the runtime trace for every sample numbered n. However, such a logging scheme consumes memory proportional to $N \log_2 M$ for ensuring reversibility from 0 to N samples, where M is the maximum number of iterations per sample. This memory is

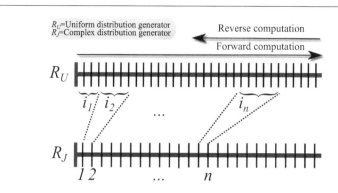

FIGURE 12.1: Variable number of uniform distribution samples used for sampling a complex distribution.

required to remember the loop count $1 \leq i_n \leq M$ for each sample numbered n, $0 \leq n \leq N$. To avoid the memory overhead, a different approach is needed that does not rely on remembering the loop counts, but can navigate back precisely to a sample in the past without using memory. Such memory-less reversible generation is described next for two common methods in the class of acceptance-rejection approaches.

The first method, namely *Upper-Bounded Rejection-Based Sampling*, relies on the availability of a special upper-bound function that covers the desired PDF everywhere. The second method, namely *Upper- and Lower-Bounded Rejection-Based Sampling*, relies on a series of upper- and lower-bound functions that cover the desired PDF from the top and bottom, respectively.

12.6.2 Upper-Bounded Rejection-Based Sampling

Consider a PDF $p(\cdot)$ for which its CDF cannot be found, or cannot be inverted, or cannot be computed inexpensively. In all these cases, it is possible to employ an acceptance-rejection approach to generate variates that satisfy $p(x)$. However, along with $p(\cdot)$, an upper-bound function $u(\cdot)$ is needed. This function is also called a *cover function* or *envelope*. The desired function $p(\cdot)$ must be covered by some scaling of the upper-bound function, that is, $p(x) \leq \alpha \cdot u(x)$ for all x, for some constant α. Also, for $u(\cdot)$, its CDF, $c_u(\cdot)$, should be known and be invertible, that is, a computable function $c_u^{-1}(\cdot)$ must be known. Note that the inversion of the CDF is used even in the traditional forward-only method to define the *forward* procedure itself.

12.6.2.1   Forward Generation

Generation of a random value conforming to $p(\cdot)$, illustrated in Figure 12.2, proceeds as follows. First, using a uniformly distributed random variable $r_1 \in [0, 1)$, the inverse CDF $c_u^{-1}(\cdot)$ of $u(\cdot)$ is used with Algorithm 12.5 to generate a random variate x_r conforming to $u(\cdot)$. Using x_r, a scaled probability value $y_u = \alpha \cdot u(x_r)$ is determined; similarly, the probability value belonging to the desired distribution is computed as $y_p = p(x_r)$. A uniformly distributed value $y_r \in [0, y_u)$ is determined using a uniformly distributed random number $r_2 \in [0, 1)$ as $y_r = r_2 \cdot y_u$. If $y_r \le y_p$, then the generated value x_r is accepted as a random sample conforming to $p(x)$. Otherwise, the candidate x_r is rejected, and the process is repeated to find another candidate. Thus, if $y_r \in [0, y_p]$, then x_r is accepted, else (i.e., $y_r \in (y_p, y_u]$) x_r is rejected. Using this procedure, on average, the accepted values of x_r follow the desired distribution $p(\cdot)$.

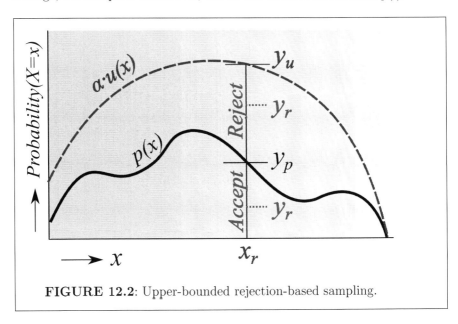

FIGURE 12.2: Upper-bounded rejection-based sampling.

12.6.2.2   Reverse Generation

The forward and reverse algorithms are shown in Algorithm 12.6. In the algorithms, the function $c_u^{-1}(\cdot)$ is the inverse CDF of $u(\cdot)$ that operates on the uniformly distributed random number r_1 to return the random sample x_r.

Two key observations about the forward algorithm are: (1) each iteration of the loop moves forward along the uniform random number stream by exactly two steps corresponding to the two invocations of $\mathcal{R}_{\mathcal{U}}()$, and (2) the iteration is exited whenever the pair of random numbers (r_1, r_2) is such that $y_r \le y_p$.

Algorithm 12.6 Reversible upper-bounded rejection-based sampling

$\mathcal{R}_\mathcal{J}()$:	$\mathcal{R}_\mathcal{J}^{-1}()$:
$N \leftarrow N + 1$	$r_2 \leftarrow \mathcal{R_U}^{-1}()\{\text{Recover recent } r_2\}$
for ever **do**	$x \leftarrow c_u^{-1}(r_2)$
$\quad r_1 \leftarrow \mathcal{R_U}()$	$\mathcal{R_U}^{-1}()\{\text{Go back past recent } r_1\}$
$\quad r_2 \leftarrow \mathcal{R_U}()$	**for** ever **do**
$\quad x_r \leftarrow c_u^{-1}(r_1)$	$\quad r_2 \leftarrow \mathcal{R_U}^{-1}()$
$\quad y_u \leftarrow \alpha \cdot u(x_r)$	$\quad r_1 \leftarrow \mathcal{R_U}^{-1}()$
$\quad y_r \leftarrow r_2 \cdot y_u$	$\quad x_r \leftarrow c_u^{-1}(r_1)$
$\quad y_p \leftarrow p(x_r)$	$\quad y_u \leftarrow \alpha \cdot u(x_r)$
$\quad$ **if** $y_r \leq y_p$ **then**	$\quad y_r \leftarrow r_2 \cdot y_u$
$\quad\quad$ **exit loop**	$\quad y_p \leftarrow p(x_r)$
$\quad$ **end if**	$\quad$ **if** $y_r \leq y_p$ **then**
end for	$\quad\quad \mathcal{R_U}()\{\text{Correct back to } r_1\}$
return x_r	$\quad\quad \mathcal{R_U}()\{\text{Correct back to } r_2\}$
	$\quad\quad$ **exit loop**
	$\quad$ **end if**
	end for
	$N \leftarrow N - 1$
	return x
(a) Forward	(b) Reverse

The reversal of the forward algorithm can be built on these two observations as follows.

When the reverse algorithm $\mathcal{R}_{\mathcal{J}}^{-1}()$ is invoked after an execution of the forward algorithm $\mathcal{R}_{\mathcal{J}}()$, it is clear that the most recent pair of uniform random numbers from $\mathcal{R}_{\mathcal{U}}()$ corresponds to the most recently accepted candidate x_r by $\mathcal{R}_{\mathcal{J}}()$. Thus, in the reverse, this value of x_r is first recovered and remembered in the variable x. Next, the uniform random number stream needs to be rewound to the correct place at which the previous $\mathcal{R}_{\mathcal{J}}()$ invocation started. This position in the stream is detected simply by searching backward in the stream for the pair of r_2 and r_1 for which the loop exit condition is satisfied. This point corresponds to the accepted value prior to the one that is being reversed. The random number pair is then moved forward once to leave it in the same place as the one at which the previous invocation of $\mathcal{R}_{\mathcal{J}}()$ started.

The total number of samples generated is tracked in the variable N, which is incremented for every forward invocation and decremented for every reversal.

12.6.2.3   Reversing the First Deviate

The reversal works perfectly until $N = 1$. Reversing the last step from $N = 1$ to $N = 0$ requires a slightly modified treatment because of the specific value of the initial seed of the random stream. The number of iterations of the **for ever** loop in the forward execution can be remembered in a single variable when $N = 0$. The random number stream can be rewound that many times (two steps back per iteration) to restore the stream to exactly the same initial value as when the stream was started. This being a minor adjustment, it is omitted for simplicity in Algorithm 12.6.

12.6.3   Upper- and Lower-Bounded Rejection-Based Sampling

A more complex variant of the rejection-based sampling is designed to reduce the computation effort for distributions whose PDF is computationally expensive. The structure of this variant, which we here refer to as the upper- and lower-bounded rejection-based sampling method, is illustrated in Figure 12.3. As with the upper-bounded rejection-based sampling method described in the preceding section, this sampling method also does not require the evaluation of the exact, closed-form inverse of the CDF of the PDF $p(\cdot)$ from which random samples are desired. Instead, it employs a sequence of progressively tighter upper $u_k(\cdot)$ and lower $l_k(\cdot)$ bound functions, $k > 0, k \to \infty$. All the upper-bound functions must be such that they completely envelop $p(\cdot)$ from above, and the lower-bound functions must envelop from below: $l_k(x) \le p(x) \le u_k(x)$ for all x and $k \ge 1$. Moreover, the bounds must get progressively tighter: $l_k(x) \le l_{k+1}(x)$ and $u_{k+1}(x) \le u_k(x)$ for all x and $k \ge 1$, and $|u_{k+j}(x) - l_{k+j}(x)| < |u_k(x) - l_k(x)|$ for some $j > 0$. The upper-bound function for $k = 0$ is a special case in that $p(x) \leq \alpha \cdot u_0(x)$ for some constant α. It also must have an additional property that it should be amenable to sampling using its inverse CDF using Algorithm 12.5.

The upper- and lower-bound functions are typically designed to be computationally easy to evaluate, relative to the computation needed to evaluate $p(\cdot)$. This helps reduce the amount of computation performed for all accepted values in the rejection method. The series of the bound functions can be separately defined for each k, or could be parameterized by k. For example, $u_k(\cdot)$ could use k as a parameter in its evaluation, as in

$$u_k(x) = U(k, x) = ke^{-\frac{k^2}{x}},$$

to systematically tighten the upper bound within an interval of interest for x. In the limit, the functions squeeze the desired function $p(\cdot)$ sufficiently tightly to approximate it to any desired accuracy. In the worst case, for a large value of k, $p(\cdot)$ itself could be used as $u_k(\cdot)$ and $l_k(\cdot)$.

12.6.3.1 Forward Generation

The pseudocode for the forward generation algorithm is listed in Algorithm 12.7. The generation relies on the inverse CDF $c_{u_0}^{-1}(r)$ to generate a sample x_r for $u_0(x)$ using a uniformly distributed random number $r_1 \in [0, 1)$ that is obtained by a call to $\mathcal{R}_\mathcal{U}()$. This gives the candidate x_r and its corresponding probability value $y_{u_0} = \alpha \cdot u_0(x_r)$. Similar to the upper-bounded method, the probability value y_r for the candidate x_r is obtained by generating yet another uniformly distributed random number r_2 and selecting $y_r \in [0, y_{u_0})$. The acceptability of x_r is tested by iteratively comparing y_r against $u_k(x)$ and $l_k(x)$ for increasing values of k. Whenever y_r exceeds $u_k(x)$, the candidate is rejected and the process proceeds to find the next candidate. Whenever y_r falls at or below $l_k(x)$, the candidate is accepted.

12.6.3.2 Reverse Generation

The challenge in reversing this method is the variable number of iterations within two nested loops that are needed for generating each sample. To go back in the stream, the number of iterations must be exactly recollected and retraced. Although it would appear as though the iteration counts must be remembered in a trace, that is in fact not needed.

The reverse algorithm without need for memory works as follows. Consider the invocation of $\mathcal{R}_\mathcal{K}^{-1}()$ after a call to $\mathcal{R}_\mathcal{K}()$. Because the successful candidate of $\mathcal{R}_\mathcal{K}()$ consumed the two most recent samples r_1 and r_2 from the uniform distribution via calls to $\mathcal{R}_\mathcal{U}()$, they can be recovered by calls to $\mathcal{R}_\mathcal{U}^{-1}()$ (first recovering r_2 and then r_1). The most recent successful candidate x_r is then reconstructed using the recovered r_2 and r_1. At this point, the uniform random number stream cannot be left in its current state because the most recent candidate may have rejected some other candidates prior to arriving at the recent candidate. Thus, the stream has to continue to be traversed

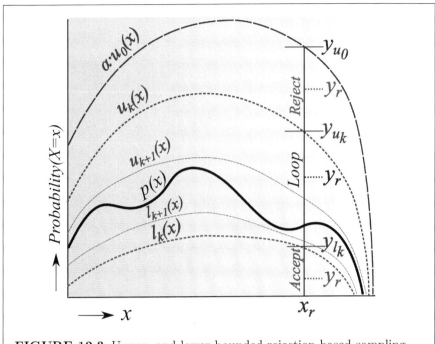

FIGURE 12.3: Upper- and lower-bounded rejection-based sampling.

backward. However, what would be the number of candidates to reject, and what would be the correct stopping point? The answer lies in the observation that the backward sequence of rejected candidates must necessarily end when a successful candidate is found. That successful candidate would correspond to the call to $\mathcal{R}_{\mathcal{K}}()$ that is prior to the most recent $\mathcal{R}_{\mathcal{K}}()$ that is being reversed. Thus, the stopping condition in backward traversal past all rejected candidates is the detection of an acceptable candidate. When this condition is met, the uniform random number stream is adjusted by moving it forward two steps, which corresponds to the detected successful candidate of the stopping condition. The complete pseudocodes for the forward and reverse generation algorithms are shown in Algorithm 12.7.

12.7   Further Reading

In general, it is very difficult, perhaps impossible, to define randomness because a deeper insight quickly touches philosophical perspectives [Bennett, 1979]. Practically speaking, there is a conventional understanding of what

Algorithm 12.7 Reversible upper- and lower-bounded rejection-based sampling

(a) Forward	(b) Reverse
$\underline{\mathcal{R}_{\mathcal{K}}()}$:	$\underline{\mathcal{R}_{\mathcal{K}}^{-1}()}$:

$\mathcal{R}_{\mathcal{K}}()$:

$N \leftarrow N + 1$
for ever **do**
  $r_1 \leftarrow \mathcal{R}_{\mathcal{U}}()$
  $r_2 \leftarrow \mathcal{R}_{\mathcal{U}}()$
  $x_r \leftarrow c_{u_0}^{-1}(r_1)$
  $y_{u_0} \leftarrow \alpha \cdot u_0(x_r)$
  $y_r \leftarrow r_2 \cdot y_{u_0}$
  $k \leftarrow 1$
  for ever **do**
    $y_{u_k} \leftarrow u_k(x_r)$
    if $y_r > y_{u_k}$ **then**
      exit inner loop {Reject}
    end if
    $y_{l_k} \leftarrow l_k(x_r)$
    if $y_r \leq y_{l_k}$ **then**
      exit outer loop {Accept}
    end if
    $k \leftarrow k + 1$ {Loop}
  end for
end for
return x_r

$\mathcal{R}_{\mathcal{K}}^{-1}()$:

$r_2 \leftarrow \mathcal{R}_{\mathcal{U}}^{-1}()$ {Recover recent r_2}
$x \leftarrow c_u^{-1}(r_2)$
$\mathcal{R}_{\mathcal{U}}^{-1}()$ {Go back past recent r_1}
for ever **do**
  $r_2 \leftarrow \mathcal{R}_{\mathcal{U}}^{-1}()$
  $r_1 \leftarrow \mathcal{R}_{\mathcal{U}}^{-1}()$
  $x_r \leftarrow c_u^{-1}(r_1)$
  $y_{u_0} \leftarrow \alpha \cdot u_0(x_r)$
  $y_r \leftarrow r_2 \cdot y_{u_0}$
  $k \leftarrow 1$
  for ever **do**
    $y_{u_k} \leftarrow u_k(x_r)$
    if $y_r > y_{u_k}$ **then**
      exit inner loop {Rejected}
    end if
    $y_{l_k} \leftarrow l_k(x_r)$
    if $y_r \leq y_{l_k}$ **then**
      $\mathcal{R}_{\mathcal{U}}()$ {Correct back to r_1}
      $\mathcal{R}_{\mathcal{U}}()$ {Correct back to r_2}
      exit outer loop {Accepted}
    end if
    $k \leftarrow k + 1$ {Looped}
  end for
end for
$N \leftarrow N - 1$
return x

would qualify a stream of numbers to be acceptable as a stream of random numbers for computational purposes. The most widely used acceptance methods are essentially constructed from a set of tests originating from expectations of coverage in various phenomena. Widely known sets of tests such as the "Diehard Battery of Tests" [Marsaglia, 1995] and "Crush" [L'Ecuyer and Simard, 2007] subject the candidate streams to a series of tests of "randomness" and provide numerical scores as bases to indicate their quality.

A truly random sequence, by definition, can never be traversed backward without recording all the bits that are part of the sequence. This is so because, otherwise, it implies that some compression of the sequence is possible, which in turn implies that some non-random portion exists within the stream, and a compression algorithm can rely on that non-random (or predictable) portion for encoding. The presence of such information violates the assumption of randomness. In other words, because no (symbolic) procedure must exist for true randomness, the sequence of numbers themselves becomes the one and only description possible for the sequence. Hence, in order to be able to reversibly traverse any sequence of perfectly random numbers, the forward sequence must necessarily be stored in a trace upon forward generation.

However, theoretically speaking, no "truly random" sequence can be identified or generated from within a closed system. In practice, one could possibly rely on highly complex physical phenomena such as the time-varying intensities of radioactivity from decaying particles, or from intergalactic radiation, and map the sequence of physical observables to numbers, to use the resulting number sequence as a random number stream. Commercial generators are capable of mapping a wide range of physical phenomena such as quantum mechanical effects to streams of bits that can pass all existing tests of randomness.

Chapter 13

Reversible Memory Allocation and Deallocation

13.1   The Problem: Reversible Dynamic Memory

Most modern programs are designed using dynamic memory management services, such as allocation and deallocation of varying sizes of memory at runtime. In traditional forward-only execution of programs, this typically translates to two basic primitives: a primitive for allocation that searches for and returns the pointer to a free region in memory that can fit the requested memory size, and another primitive for deallocation of previously allocated memory regions for returning into the free pool of memory for later reuse. When these two primitives are to be utilized inside a reversible program, however, new issues arise that are simply absent in a forward-only mode. An allocation must be undone by freeing the allocated memory if the allocation operation itself is reversed. Similarly, a deallocation operation must *not* immediately release the memory because the deallocation itself may be reversed. Otherwise, the memory incorrectly released may be reused by another allocation, which will result in an undesirable situation in which two logically distinct objects occupy the same memory space. Moreover, the dynamic memory management interface should be extended so that it is dynamically provided information about guarantees that a particular allocation is not tentative, and about situations when a deallocation is in fact correct (i.e., the deallocation will not be reversed).

The unique considerations in adding reversibility to dynamic memory management are satisfied using the Forward-Reverse-Commit (FRC) paradigm of reversible execution. The basic idea is that the forward operations will include the usual forward-only actions, augmented with sufficient information to deal with the special cases introduced due to reversibility. The reverse operations will make the necessary changes to undo the side effects, while the commit stage is used to finalize the allocation or deallocation. To use this paradigm, the operations in the product $\{allocate, deallocate\} \times \{forward, reverse, commit\}$ must be defined. In this space, two solutions are possible: one is a quick and dirty solution that provides a degenerate solution that is very easily implemented, while another is a more sophisticated solution that implements reversibility semantics both correctly and efficiently, but which is more complex to implement.

In what follows, a void operation, also called a *no-op*, is denoted by $\square$.

13.2   A Simple Solution with Poor Memory Utilization

One of the easiest solutions is to perform allocation as usual, but always forego deallocation. Memory allocation of traditional forward-only execution can be retained unmodified in the reversible execution setting, and the deallocation operation is made a void operation. No memory is ever freed.

This scheme would in fact work adequately in the sense that it retains all memory semantics as forward-only execution (memory allocated dynamically from the free pool, and memory is never accidentally shared by two or more distinct objects), and thus will result in correct reversal of any forward execution. However, the biggest drawback is its poor memory utilization—a leakage of memory occurs when memory that is allocated in a forward execution is later reversed. When a number of iterations of forward- and reverse execution paths are traversed, they leave behind memory that is marked occupied but not used. Although the application assumes that a memory block is freed, that block is never returned to the pool, and is hence irretrievably lost. Available free memory becomes depleted over the length of execution.

TABLE 13.1: Simple Procedures for Dynamic Memory Management

Operation P	Traditional Forward-only $\overline{F}(P)$	Reversible $FRC(P)$		
		Forward $F(P)$	Reverse $R(F(P))$	Commit $C(F(P))$
Allocation	m=malloc()	m=malloc()	$\square$	$\square$
Deallocation	free(m)	$\square$	$\square$	$\square$

Thus, the simple solution of making deallocation a no-op gives correct execution but with poor memory utilization. In situations demanding a quick implementation of reversibility support for dynamic memory, this solution can be adopted as a first cut of a working version. Later, however, when the problem of poor utilization needs to be remedied, a more comprehensive, memory-efficient solution can be adopted, as described next.

13.3 A Memory-Efficient Solution

The key to a reversible and *efficient* solution to dynamic memory management is the introduction of a memory operation stack, which we will abbreviate as *mstack*. This is a stack of pointers to allocated memory regions that the memory manager needs to remember from the forward execution stages to recollect and use in the reverse execution or commit stages. Because reversals follow last-in-first-out order, the stack data structure correctly recovers the most recently used pointer simply by popping the top of the stack. With the help of the *mstack*, it then becomes a matter of defining the memory operations along with stack operations for each of the six elements in FRC-based dynamic memory allocation. The templates of their basic operation are listed in Table 13.2, with details such as memory sizes omitted. The code m=malloc() represents a call to a conventional (forward-only) allocator that returns a pointer m to a free memory region that fits the requested memory size, and free(m) represents a call to a conventional (forward-only) allocator that marks the region pointed to by m as free for reuse in a later allocation. The operation push(m) pushes the value of the pointer m to *mstack* while pop() returns the pointer that is currently on the top of the stack.

- **Forward-Allocation** In the forward stage, memory is allocated using any conventional (forward-only) allocator, which we represent by malloc() that returns a pointer m to a free region in memory that fits the requested memory size. Before this pointer is returned to the requestor, it is remembered by pushing it onto the *mstack* using push(m).

- **Reverse-Allocation** When a previous allocation is reversed, we can be sure that the pointer to the previously allocated region would be found on the top of the stack. This pointer m is retrieved using m=pop() and it is freed using the conventional (forward-only) allocator represented by free(m). This effectively nullifies the previous allocation, with respect to both the memory subsystem as well as the *mstack*.

- **Commit-Allocation** This indicates that the previous memory allocation can be safely assumed to be lasting, and hence the pointer can be forgotten with respect to reversibility. In other words, the previous allocation can be viewed as equivalent to the forward-only counterpart. Hence, it is sufficient to simply remove the pointer from the *mstack*, which is accomplished via pop(). Because the requestor possesses a copy of the pointer to the allocated memory block, the pop operation can ignore and discard the popped top pointer.

- **Forward-Deallocation** In the forward stage of deallocation, the memory being deallocated cannot be marked as free right away because the deallocation can later be reversed. However, if it is *not* reversed and in fact it is later committed, we will need to recollect that this block was marked for deallocation and its deallocation must be effected, so that it can be reused in a later allocation. Thus, to enable use in either later reversal or commitment, the pointer is remembered by pushing it on to the *mstack*. No additional processing is done on the pointer.

■ **Reverse-Deallocation** This marks that a previous deallocation is being undone. Because we had only stored the pointer on the *stack*, it is necessary and sufficient to simply pop that pointer off the stack – the deallocation will be completely reversed by this act. Thus, pop() is used to remove the pointer currently on the top of the stack and that pointer is forgotten.

■ **Commit-Deallocation** Because this is an indication that the deallocation will be permanent, it needs to be effected immediately. The pointer in question is retrieved by popping the top of the stack, and the pointer is passed to the conventional (forward-only) allocator to mark it for reuse. This completes the full cycle of reversible memory allocation and deallocation that is subject to reversal at either stage.

TABLE 13.2: Reversible Procedures for Dynamic Memory Management

Operation	Traditional	Reversible		
P	Forward-only $\overline{F}(P)$	Forward $F(P)$	Reverse $R(F(P))$	Commit $C(F(P))$
Allocation	m=malloc()	m=malloc() push(m)	m=pop() free(m)	pop()
Deallocation	free(m)	push(m)	pop()	m=pop() free(m)

13.3.1   Verification of Correctness of Allocation

We can verify that the semantics of the original forward-only execution $\overline{F}$(Allocation) is indeed maintained in the reversible execution FRC(Allocation). Because the operation m=malloc() is nullified by free(m), we can remove both of them, giving a no-op: $\boxed{\text{m=malloc(); free(m)}} = \square$.

Similarly, Because a push(m) is nullified by a pop(), $\boxed{\text{push(m); pop()}} = \square$.

Applying the nullification equations to the terms in the reversible equation

$FRC(P)$ for $P =$ Allocation, we obtain the following.

$$\overline{F}(P) = [F(P) \rightsquigarrow R(F(P))]^* \rightsquigarrow F(P) \rightsquigarrow C(F(P))$$

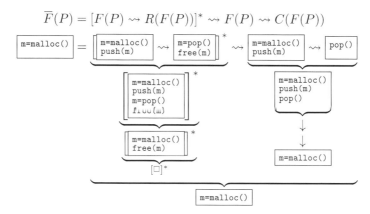

Thus, the net effect of reversible allocation is the same as that of traditional (forward-only) allocation operation.

13.3.2 Verification of Correctness of Deallocation

Similarly, applying the nullification equations to the terms in the reversible equation $FRC(P)$ for $P =$ Deallocation, we obtain the following.

$$\overline{F}(P) = [F(P) \rightsquigarrow R(F(P))]^* \rightsquigarrow F(P) \rightsquigarrow C(F(P))$$

Thus, the net effect of reversible deallocation is the same as that of traditional (forward-only) deallocation operation.

Chapter 14

Reversible Numerical Computation

14.1   Software and Hardware Views

The issue of reversibility of numerical computation can be approached from two perspectives: the software point of view and the hardware point of view. Reversibility in software is concerned with the ability to execute a sequence of arithmetic operations in both directions of the sequence: using normal operators in the forward direction, and using inverse operators in the reverse direction to recover operands that are lost or overwritten in the forward direction. Reversibility in the software view is concerned about whether and how the loss of some operand values can be reversed by *uncomputing* the operator using the results of the operator and the rest of the operands that are not lost. Reversibility of arithmetic in the hardware is concerned with the extent to which the arithmetic logic circuitry can be made reversible, typically by employing reversible gates. The hardware view is also concerned with whether and how a numerical algorithm could be expressed using only reversible *instructions* at the machine code level.

In this chapter, some of the important issues and considerations are identified in adding reversibility to numerical computation in both hardware and software points of view. A few solution approaches are examined, and some of the outstanding challenges are outlined.

14.2   Sources of Irreversibility in Software

What contributes to irreversibility in a sequence of arithmetic computations? The following are some of the factors:

S1 **Fundamental Loss** A fundamental loss of information may occur due to the nature of the operator itself, which necessarily introduces ambiguity. For example, the sign of a number is lost if only its modulus or square is remembered. Similar loss occurs with operations such as trigonometric functions; for example, the period information of an angle is lost if only its sine value is remembered. In special cases of operand values, information may be destroyed by the computation of the operator itself, making reversal impossible unless the destroyed information is saved by copying to another variable. The most notable example of such loss is multiplication by zero.

S2 **Memory Overwriting** Information may be lost when variables containing the operand values are overwritten with other values. The memory location of one or more operands may be overwritten by the operator's result or by other operations executed later in the program.

S3 **One-to-Many Mapping** A one-to-many mapping may arise in the implementation of operations, which makes the operations irreversible. For example, round-off in floating point computation necessarily introduces ambiguity with respect to one or more bits in the operand(s).

S4 **Order of Operations** For a sequence of operations, if the order of operations performed in the forward direction is altered (even for a single item) in the reverse direction, the computation can potentially lead to different results. This is due to lack of commutativity and associativity guarantees of many computational operations, notable of which are the non-associative and non-commutative aspects of floating point arithmetic.

S5 **Sticky Errors** In a sequence of operations, if an error condition occurs at one point, the error can become sticky from that point forward, even if the operations that follow are otherwise reversible. This makes the results undefined and consequently irreversible. Examples of common sticky error conditions are an attempt to divide by zero or to compute the square root of a negative number. Special numerical codes such as NaN (not-a-number) that arise in such sticky conditions can render an entire series of operations irreversible.

S6 **Hardware Non-Determinism** Non-deterministic effects from computational hardware in the computation of some functions (such as numerical approximations to transcendental functions) may seep into the values of results. The main problem with such non-deterministic noise is that the random values of certain bits will be different in forward and reverse modes, which results in differences between operands used in forward mode and the operand values recovered in reverse mode.

14.3   Considerations in Adding Reversibility

Some of the considerations in adding reversibility to computer arithmetic operations are listed in Table 14.1. A major consideration regards the type of numbers to be supported—integer data types, fixed precision (also known as fixed point) data types, or floating point data types. The way reversibility is added is significantly dependent on the choice of data types. In general, operators with only integers or fixed point operands suffer from fewer sources of irreversibility, whereas floating point requires more research and development, especially when backward compatibility is desired. Operations can be defined to follow modular arithmetic rules, or may assume variable precision. Implementations may be realized in software or manifested in specially optimized hardware. Reversible arithmetic solutions may be sought in a classical computing setting or be investigated in the context of quantum computing with reversible gates from the outset. Hardware instruction sets and software interfaces may be sought either for entirely new versions that are reversible by design or for accommodating and extending currently used interfaces. In the short term, one is faced with the question of how to reuse the immense amount of existing forward-only technology for arithmetic. In a medium-term outlook, it is useful to understand how to mix conventional, optimized forward-only arithmetic with newer, reversible arithmetic. In the longer term, the ideal would be to move to a fundamentally reversible arithmetic interface and implementation on all computing platforms.

TABLE 14.1: Considerations in Adding Reversibility to Computer Arithmetic Operations

Consideration	Notes
Integer *vs* Floating-point Numbers	Whether numerical computation is limited to integers or if real numbers are also supported
Normal *vs* Modular Arithmetic	Whether or not all computations are restricted to modular arithmetic
Hardware *vs* Software Implementation	Layer at which the implementation of reversible arithmetic is realized
Classical *vs* Quantum Computing	Computing paradigm on which reversible arithmetic is applied
Backward-Compatible *vs* New Interface	Whether numerical computation interfaces are permitted to be developed anew, and/or if backward compatibility is to be supported for existing forward-only interfaces

14.4   Defining Reversibility of Numerical Computation

While one can easily arrive at an intuitive notion of reversal of numerical computation based on basic mathematics, a generally open view of reversal of any mathematical operation goes beyond computational aspects. Instead, to restrict the discussion to computational issues that arise in conventional uses of computer software and hardware, it is useful to have a set of working definitions that delimits the scope of reversibility. In this context, reversibility can be examined of (1) numerical computations as *individual* operators and their *sequences* defined at the software level, and (2) operations realized at the hardware *circuit* level.

14.4.1   Software-Level Operator Reversal

At the software-level, consider the following setting for the execution of an arithmetic operation

$$\odot : (o_1, \ldots, o_n) \to O$$

defined by an operator $\odot$ with numerical operands $o_1, \ldots, o_n$ giving the result O. All values are assumed to be of finite precision, but the precision can be different for different operands. The result O is often a single value $\hat{o}$ but could potentially be a composite value $O = (\hat{o}_1, \ldots, \hat{o}_m)$. After the operator is applied and the result is recorded, one or more operands may be "lost" or "forgotten" or overwritten by the program. The mutual order of operands is also similarly susceptible to loss.

In this setting of potential loss of operands or their relative order, we define reversibility of the operator in software as the capability to fully and correctly recover all the original operand values *and* the relative order of the operands *after* the operator has been applied. To facilitate such recovery, the original operator $\odot$ may be modified into a new operator $\odot^{+1}$ to allow for such recovery of operands while retaining the original semantics of computing O from the operands just as in the original. The recovery procedure $\odot^{-1}$ is called the inverse operator of $\odot^{+1}$. Thus, $\odot$ is the original operator, $\odot^{+1}$ is the modified forward operator, and $\odot^{-1}$ is the inverse of $\odot^{+1}$.

14.4.2   Operators with Constants

If any of the operands is a constant, it can be simply treated as a special case in which the specific operand(s) cannot be forgotten, lost, or overwritten. For example, in an operation of the form

$$+ : (\mathtt{A}, c) \to \overline{\mathtt{A} + c}, \text{ or, more commonly written as } \mathtt{A} \leftarrow \mathtt{A} + c$$

on a variable $\mathtt{A}$ with a constant c, $\mathtt{A}$ is overwritten, but the operand c cannot be lost because c is a constant. Reversibility is relatively straightforward in such cases. A commonly used special case of constants as operands is the *increment* operator on a variable. In this case, $c = 1$, giving $\mathtt{A} \leftarrow \mathtt{A+1}$, or simply $\mathtt{A++}$, whose inverse is $\mathtt{A} \leftarrow \mathtt{A-1}$, or simply $\mathtt{--A}$.

14.4.3   Operation Sequence Reversal

In a sequence of operations, even if a single operation is irreversible, the reversibility of the sequence is affected in the reverse mode, starting from that irreversible operation to the beginning of sequence. In other words, in the sequence of operations $\odot_1, \ldots, \odot_k$, where $\odot_i : (o_{i_1}, \ldots, o_{i_{n_i}}) \to (\hat{o}_{i_1}, \ldots, \hat{o}_{i_{m_i}})$, if even one operation $\odot_i$ does not restore its operands to their correct previous values, the reversal of all operators $\odot_j, 1 \leq j < i$ will be affected. Thus, the issue of reversibility is of an "all-or-nothing" nature: either *every* type of arithmetic operation is ensured to be reversible or *no guarantee* of arithmetic reversibility is provided by the computing system.

14.4.4   Hardware Circuit-Level Reversal

Reversibility at the level of hardware circuits can be defined as the ability to implement one or more basic mathematical operations as a reversible circuit that not only computes the desired output bits from the input bits but can also recover the input bits from the output bits. The same can also be viewed as the problem of defining a reversible truth table inside which the mapping of the desired operation is embedded. For example, a reversible adder is a circuit that provides the sum of two input bit vectors but preserves sufficient information in the output such that the original operands can be recovered when driven backward. Such a reversible circuit may be built using a set of reversible gates. Moreover, the issue of reversibility may be considered in the context of classical computing as well as quantum computing hardware systems. Circuits may be defined on the basis of each individual operator separately, or may be defined as a holistic Arithmetic Logic Unit (ALU) for an aggregate set of operators.

14.5 Reversal of Basic Arithmetic Operations in Software

The easiest reversal method, which can always be used as a fall-back approach, is to save a copy of the operand(s) being overwritten or forgotten. All additional improvement of reversal is an attempt to avoid such saving of one or more variables, or to reduce the overall number of bits to be saved, in exchange for recovery of values via computation. The fundamental premise underlying the improvement is that computation-based recovery is preferable to saving copies of values due to memory considerations. However, logging some values to a history trace may be unavoidable in general, and the use of a history trace can reduce the computation speed relative to forward-only computing. The amount of data to be saved to reverse an operation varies with the data type of the operands. In general, operations with integer operands are relatively easier to optimize than those with floating point operands.

14.5.1 Illustration of Basic Approach

Consider an instruction of the form

$$C \leftarrow A*B$$

in which the product of two operands, variables A and B, is stored in a variable C, after which one of the operands, A or B, is forgotten. Another variant of the same instruction is of the form

$$(A,D) \leftarrow A*B$$

in which a variable A is overwritten with a part of the product of itself and another variable B. In the preceding instructions, if A, B, and D are w bits wide, then C must be at least $2w$ bits wide.

To make any such operation reversible, it is replaced with a different type of instruction that, by design, does not incur loss of information. For example, the operation C←A*B is replaced with an instruction of the form

$$(C,A,T) \leftarrow A*B$$

in which the product of two numbers A and B is computed and the result is stored in the variable C whose bit width is equal to sum of bit widths of A and B, and the value of A or B (whichever is non zero) is remembered in A. T is a bit variable that remembers if A was equal to zero. This organization allows the value of B to be forgotten after the operation, yet can be perfectly recovered using C, A, and T. Let the size of A and B be w bits each; the size of C is $2w$ bits. Compared to the memory consumed by an implementation that saves all operands that are $2w$ bits (w bits each for A and B), the reversible version (C,A,T)←A*B consumes only $w + 1$ bits (1 bit for T and w bits for A). Note that $w + 1$ is the minimum number of bits needed to reverse the product operation. The only rare exception is the case in which the program somehow can guarantee that A and B are *always* prime numbers, in which case the minimum memory size for reversal is only one bit (to only remember the assignation, namely whether the smaller of the two prime numbers was A or B). Analogous constructs and semantics can also be defined for other operations such as division.

Alternatively, the operations can be redefined to be reversible without history. To aid in such redefinition, the internal representation of the operands and results of operations must be set up to store information in a lossless fashion.

We will now examine the reversal of conventional arithmetic operations, focusing on those with integer operands, and discuss the issue of reversing operations on floating point operands. Following that, we will study a new, alternative framework for reversible integer operations that first defines a new bit format for internal representation of integers, and then defines reversible integer arithmetic operations that do not result in any history generation.

14.5.2 Integer Operands

A list of conventional software-level operators is shown in Table 14.2, which are designed for use in typical forward-only codes. In many cases, the operator may be defined (separately) into different versions for integer and real-valued variables. In this section, we will consider reversal of these operations with integer operands only.

For the reversal algorithms, we assume the maintenance of a history tape operated in a last-in-first-out (stack) order, simiilar to that described in Section 10.2.3. The routine named **save(v)** stores a copy of the variable v on the tape, while the routine named **restore(v)** overwrites the variable v with the value popped from the end of the tape. The variable being saved could be of any size, as small as a single bit or several bytes long.

TABLE 14.2: Computer Arithmetic Operations Considered for Reversibility

Operation	$\odot$	Syntax	Post-Operational Effects
Add	+	C ← A + B	A or B forgotten
Accumulate	+=	A ← A + B	A overwritten
Subtract	-	C ← A - B	A or B forgotten
Diminish	-=	A ← A - B	A overwritten
Complement	-=-	B ← A - B	B overwritten
Multiply	*	C ← A * B	A or B forgotten
Semi-scale	*=	(A,D) ← A * B	A overwritten
Scale	*=*	(A,B) ← A * B	A and B overwritten
Divide	/	C ← A / B	A or B forgotten
Shrink	/=	A ← A / B	A overwritten
Factor	/=/	B ← A / B	B overwritten
Remainder	%	C ← A % B	A or B forgotten
Modulo	%=	A ← A % B	A overwritten
Base	%=%	B ← A % B	B overwritten
Separate	@	(C,D) ← A @ B	A or B forgotten
Semi-partition	@=	(A,D) ← A @ B	A overwritten
Partition	@=@	(A,B) ← A @ B	A and B overwritten

   $\boxed{+}$ $\boxed{\text{C←A+B}}$ The **Add** operation assigns the sum of A and B to C. If A is forgotten after this operation, it is easily recovered as A←C-B (**Subtract**). Recovery of B from A is also symmetric.

   $\boxed{+=}$ $\boxed{\text{A←A+B}}$ The **Accumulate** operation increases the value of A by B. It is easy to recover the previous value of A via A←A-B (**Diminish**).

   $\boxed{-}$ $\boxed{\text{C←A-B}}$ The **Subtract** operation assigns the difference of A and B to C. If A is forgotten after this operation, it is easily recovered as A←C+B (**Add**). Similarly, B can be recovered as B←A-C.

   $\boxed{-=}$ $\boxed{\text{A←A-B}}$ The **Diminish** operation diminishes the value of A by B. It is easy to recover the previous value of A via A←A+B (**Accumulate**).

   $\boxed{-=-}$ $\boxed{\text{B←A-B}}$ The **Complement** operation replaces B by its complement with respect to A. It is easy to recover the previous value of B via B←A-B (self-inverse).

   $\boxed{*}$ $\boxed{\text{C←A*B}}$ The **Multiply** operation assigns the product of A and B to C. The bit-width of C must be twice that of A and B. Let us assume that B is forgotten after this operation. Then, B can be recovered as B←C/A. However, when A*B equals zero, additional information is needed to properly reverse the operation to recover B because B could have been non-zero, which cannot be recovered from C alone.

Thus, in general, if A and B are each w bits wide, and C is $2w$ bits wide, a total of $3w + 1$ bits are needed to reverse the operation: the $2w$ bits of C, the w bits of a non-zero operand (either A or B, whichever is non-zero), and 1 bit to remember which of the two operands A or B was zero in case only one of them is zero. There is also one special case in which $2w$ bits are sufficient—this is when A and B are prime numbers, allowing C to be uniquely partitioned as a product of two primes, but this is difficult to utilize in memory-saving schemes. Thus, there are four cases to be considered:

1. A=0, B≠0, or
2. A≠0, B=0, or
3. A=0, B=0, or
4. A is prime, B is prime.

One bit can distinguish between the first two cases. The third case is easily detected when C=0 and the remembered value (A or B) itself is also zero. The fourth case is a rare case, in which no operand needs to be saved, but still one bit is needed to store the order of the operands (was the smaller prime A or B?). Hence, a minimum of $2w + 1$ bits and a maximum of $3w + 1$ bits are needed to reverse this multiplication operation. In fact, the minimum can be reduced to just 1 bit, which remembers whether A and B were *both* equal to zero; however, this adds one bit to the maximum, increasing it to $3w + 2$. Because the situation of both A and B being equal to zero is rare, we will ignore this potential reduction of the minimum number of bits, and assume that one value (either A or B) is always remembered.

Let T be the 1-bit variable that is pushed onto the history stack for each execution of the multiplication operation. The multiplication operation is made reversible as shown in Algorithm 14.1, assuming C, A, and T are available after the forward multiplication, but B needs to be recovered. Because A and B are symmetric, the same procedures can be used by swapping the identities of A and B.

$\boxed{*=}$ $\boxed{(\mathtt{A},\mathtt{D})\leftarrow\mathtt{A}*\mathtt{B}}$ The **Semi-scale** operation assigns the product of A and B to the aggregate variable formed by juxtaposing A with another variable D. Note that the old value of A is lost as it is overwritten by the high bits of the A*B product. The bit widths of A, B, and D are the same, say, w bits each. Thus, the aggregate (A,D) denotes the $2w$ bit-wide double precision product. In order to be able to reverse this operation, we must either assume that B is not forgotten after this operation, or, if that cannot be ensured, that B is saved on the history stack and restored just before reversal. Similar to the **Multiply** operation, special handling

Algorithm 14.1 Reversal of the **Multiply** operation

Forward-only	Reversible	
	Modified Forward	Reverse

C←A*B	T←0
	if A=0 then
	    T←1
	    A↔B {Swap}
	end if
	C←A*B
	save(T)
		restore(T)
		B←0
		if A≠0 then
		    B←C/A
		    if T=1 then
		        B↔A {Swap}
		    end if
		end if

	History size: 1 bit	

is needed to take care of the case when A*B equals zero; additional information is needed to properly reverse the operation to recover A because A could have been non-zero, which cannot be recovered from A,D alone if B is zero.

Thus, in general, a total of $3w+1$ bits are needed to reverse the operation: the $2w$ bits of (A,D), the w bits of a non-zero operand (either A or B, whichever was non-zero), and a 1-bit variable T to remember which of the two operands A or B was zero in case only one of them was zero.

Algorithm 14.2 Reversal of the **Semi-Scale** operation

Forward-only	Reversible	
	Modified Forward	Reverse

(A,D)←A*B	T←0
	if B=0 then
	    T←1
	    B↔A {Swap}
	end if
	(A,D)←A*B
	save(T)
		restore(T)
		A←0
		if B≠0 then
		    A←(A,D)/B
		    if T=1 then
		        A↔B {Swap}
		    end if
		end if

	History size: 1 bit	

Analogous to those of **Multiply**, the forward and reverse operators are as shown in Algorithm 14.2. The algorithm assumes that (A,D), B, and T are available after the forward multiplication, and A needs to be recovered. The operation A←(A,D)/B denotes the integer division in which the aggregate integer (A,D) is the dividend and B is the divisor.

If B is forgotten instead of A, the same procedures can be used by swapping the identities of A and B, as A and B are symmetric.

| *=* | (A,B)←A*B | The **Scale** operation stores the product of A and B in the aggregate variable formed by juxtaposing A with B. Thus, the old value of A is overwritten by the high bits, and D is overwritten by the low bits of the A*B product. This operation can be undone in a manner similar to **Multiply** and **Semi-scale** if the pre-assignment value of A or B (whichever is non-zero) is saved to the history stack in the forward direction and restored from the tape in the reverse direction. A 1-bit value T is used to remember if A and B were swapped before saving to history.

| / | C←A/B | The **Divide** operation replaces C with the quotient of A/B. Reversal of this operation implies the recovery of the operand that is forgotten after the operation: recovery of A if A is forgotten, and recovery of B if B is forgotten. Two cases hinder the reversal of this operation: when the dividend is zero and when the divisor is zero. When the dividend is zero, information about the divisor will not be preserved in the quotient; and when the divisor is zero, the dividend is similarly lost. These cases are solved by writing a few bits of information to the history tape.

The forward and reverse procedures are shown in Algorithm 14.3. When the divisor B is forgotten, it is sufficient to write 2 bits of information to the tape to deal with all cases. A bit TB is used to remember if B is zero, and another bit TB is used to remember if A is zero. In the special case when A=0 and B≠0 (indicated by the condition TA=1 and TB=0), the value of B is saved in A.

| /= | A←A/B | The **Shrink** operation overwrites A with the quotient of A/B. This operation is difficult to reverse without saving w bits in the worst case, irrespective of any scheme used. Thus, if B is forgotten after the operation, it must be saved to the history in the forward execution. Even if B is not lost after the operation, the remainder value A%B needs to be remembered to properly recover the pre-operation value of A. Because the bit width of the remainder can be as large as the value of A itself, it is easier to simply save the pre-operation value of A itself to the history. Thus, the solution is to save the value that is being forgotten; this takes w bits of history.

| /=/ | B←A/B | The **Factor** operation overwrites B with the quotient of A/B. This operation is reversed differently based on whether A is forgotten or not after the operation. If A is forgotten, the information in B is inadequate to recover A, and hence A needs to be saved to the history.

Algorithm 14.3 Reversal of the **Divide** operation when A or B is forgotten

Forward-only
C←A/B

Reversible	
A is forgotten after Forward and restored in Reverse	
Forward	**Reverse**

if B=0 then
　T←1
　C←A
else
　T←0
　C←A/B
　D←A%B
　save(D)
end if
save(T)
	restore(T)
	if T=1 then
	　A←C
	else
	　restore(D)
	　A←C*B+D
	end if

History size: Minimum 1 bit and maximum $w+1$ bits	

Reversible	
B is forgotten after Forward and restored in Reverse	
Forward	**Reverse**

if B=0 then
　TB←1
　C←*Undef*
else
　TB←0
　C←A/B
　TA←0
　if　　A=0
　then
　　TA←1
　　A←B
　end if
end if
save(TB)
save(TA)
	restore(TA)
	restore(TB)
	B←0
	if TB=0 then
	　if　　TA=1
	　then
	　　B←A
	　else
	　　B←A/C
	　end if
	end if

History size: 2 bits	

However, if A is not forgotten, B can be recovered by simply invoking the same operation (self-inverse), as B←A/B. The only special case to be addressed is when A=0, in which case B must be logged to the history.

| % | C←A%B | The **Remainder** operation assigns to C the remainder from the division of A by B. This operation can be reversed using an approach similar to that for **Divide**.

| %= | A←A%B | The **Modulo** operation overwrites A with the remainder from the division of A by B. This operation can be reversed using an approach similar to that for **Shrink**.

| %=% | B←A%B | The **Base** operation overwrites B with the remainder from the division of A by B. It can be reversed using an approach similar to that for **Factor**.

| @ | (C,D)←A@B | The **Separate** operation computes A/B and assigns the quotient to C and remainder to D. Few programming environments provide such an operation, but it is more readily amenable to reversal. After the operation, if A is forgotten, it is easily reconstructed as C*B+D. If B is forgotten, it can be reconstructed as A/C. The special case of A=0 can be handled by overwriting A with B (i.e., remembering B in A), and the special case of B=0 can be handled by copying A into D.

| @= | (A,D)←A@B | The operation of **Semi-partition** is similar to that of the **Separate** operation.

| @=@ | (A,B)←A@B | The operation of **Partition** is similar to that of the **Separate** operation, except that A and B are overwritten with the quotient and remainder of A/B.

14.5.3 Floating Point Operands

Floating point arithmetic in general is irreversible when considered at the level of each individual operator or as a sequence of operations. For example, all basic IEEE floating point operations of addition, subtraction, multiplication, division, remainder, and square root [Goldberg, 1991] exhibit irreversibility under various conditions. There is little reversibility at the hardware level either. The sources of irreversibility in floating point come in all forms.

In a fundamental viewpoint, floating point conventions such as the IEEE 754 standard and the IEEE 854 standards are essentially a set of compromises with which we have evolved to perform computations in conventional computing. Hardware technologies impose the realities of limited precision, and computing needs impose the realities of speed expectations and memory limits. Taking them into account, a set of trade-offs have evolved over time to reconcile fundamental mismatches such as infinite-to-finite precision for the real number line and the resultant travesties such as anti-commutative and anti-associative properties. There is an important aspect to note about such an evolution in numerical computation: it may not be justifiable to attach an undue amount of sanctity to the state of affairs in floating point computation as we have it now. If the consideration of reversibility is added to the set of criteria to redefine floating point computation, the overall solution may look different from its current form, but it would not affect the essential nature of floating point arithmetic, namely that it is nothing but a *compromise* in mapping the infinite precision of real numbers to the limited precision of floating point representations.

It is possible to implement in software all floating point arithmetic operations in a reversible way using integer arithmetic that can be implemented reversibly using modulo arithmetic. The challenges in this approach, however, are twofold: (1) it entails a large amount of implementation effort, due to the need to re-implement most common operations albeit reversibly (the literature is vast with respect to optimized implementations that are irreversible); and (2) it suffers from a slow-down due to implementation in software, as opposed to the extremely highly optimized realizations of conventional arithmetic in hardware (floating point units). Added to this is the need to restructure the software to conform to new interfaces and software reorganizations needed to conform to the new reversible implementation semantics. The area of reversible arithmetic is wide open for new results and advances. One method is to define new floating point algorithms with reversibility semantics (analogous to the IEEE floating point representation and operation standards), and begin the process of developing optimized algorithms to implement the semantics using newly defined and developed hardware support. Another method, perhaps pursuable concurrently with the previously mentioned method, is to find ways to utilize a subset of currently hardware-optimized operations as building blocks for reversible versions of higher-level operations.

Techniques such as program synthesis [Jha, 2011] may be applied to automatically convert floating point-based codes to equivalent fixed point-based codes that are guaranteed to compute the same answers. Also, certain application-specific customized solutions of reversible floating point arithmetic [Levesque and Verlet, 1993, Bowers et al., 2006] have been proposed and applied. For example, all operations that are subject to reversal are first converted from floating point to fixed point format, reversible fixed point arithmetic is performed, and finally the results are placed back in floating point format; the order of operations is also ensured to be deterministically reproducible in the reverse mode [Bowers et al., 2006].

14.6 Alternative Integer Framework for Reversibility

With the conventional forward-only operations and the associated internal representation of operands, some operations are easy to reverse but others do not seem natural. The reversal of operations based on addition and subtraction are relatively straightforward, but the reversal of others based on multiplication and division are somewhat awkward and inelegant, giving rise to a range of operators that differ slightly from each other. The definition of an alternative view of the internal representation of integers is intended to generalize and streamline the reversibility of the basic arithmetic operations.

The bit representation A of a w-bit positive integer $\mathcal{A} = \text{Value}(A)$ is typically a simple sequence of bits

$$A = \boxed{a_{w-1}} \boxed{a_{w-2}} \ldots \boxed{a_1} \boxed{a_0}$$

with value

$$\mathcal{A} = a_{w-1}2^{w-1} + a_{w-2}2^{w-2} + \ldots + a_1 2^1 + a_0 2^0.$$

While such a representation does not pose any problems for the reversal of addition and subtraction operations, it is not readily amenable to reversal of operations that compute division and remainder. In particular, conventional arithmetic becomes irreversible due to loss of information in (1) conventional division or remainder operations, and (2) scaling a variable by a factor equal to zero. We will now look at a generalization (or relaxation) of the conventional representation to overcome the inherent irreversibility of conventional arithmetic, and is better suited for reversible arithmetic operations, including division.

14.6.1 Internal Representation

The alternative internal representation of an integer variable is shown in Figure 14.1. For every w-bit integer $\mathcal{A}$ conventionally represented in a w-bit variable A, a new "control" field L that is l bits wide is added to the internal representation, where $l = \lceil \log_2 (w+1) \rceil$. This increases the total bit length of the bit representation of $\mathcal{A}$ from w bits to $w + l$. The control field is used to store a "quotient offset" value q that divides the w bits of A into two parts. The first part is an *encoded quotient* number $\hat{Q}$ stored in the lower q bits $A[0, q-1]$. The second part is a *remainder* R stored in the upper $r = w - q$ bits $A[q, w-1]$.

For example, a 32-bit integer is represented using $32 + \lceil \log_2 33 \rceil = 38$ bits, and a 64-bit integer is represented using $64 + \lceil \log_2 65 \rceil = 71$ bits.

The major difference between conventional representation using w bits and this new alternative representation using $w + l$ bits is the presence of a new,

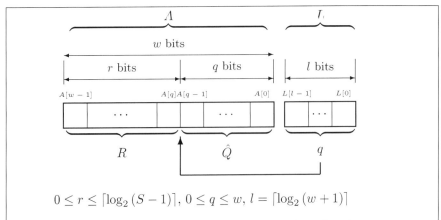

$$0 \le r \le \lceil \log_2 (S-1) \rceil,\ 0 \le q \le w,\ l = \lceil \log_2 (w+1) \rceil$$

FIGURE 14.1: Alternative internal representation of an integer amenable to reversible arithmetic.

implicitly underlying "divisor" S. For a specific integer S, $1 \le S \le 2^w$, the following relations hold among the various values in the new representation:

$$0 \le R < S, \quad R = \begin{cases} 0 & \text{if } q = w \\ A[q, w-1] & \text{otherwise } (0 \le q < w), \end{cases}$$

$$0 \le Q < \left\lfloor \frac{2^w - 1}{S} \right\rfloor, \quad \hat{Q} = \begin{cases} 0 & \text{if } q = 0 \\ A[0, q-1] & \text{otherwise } (0 < q \le w), \end{cases}$$

$$Q = 2^q - 1 + \hat{Q},$$

$$\mathcal{Q}(A) = Q,\ \mathcal{R}(A) = R,\ \text{and } \mathcal{A} = \text{Value}(A) = \boxed{A}_S = QS + R\ .$$

The special value of $q = w$ corresponds to the case when all bits of A are used to represent Q, which implies $R = 0$. Similarly, the special value of $q = 0$ corresponds to the case when all bits of A are used to represent R, which implies $Q = 0$.

14.6.2 Encoding Certain Error Conditions

In a separate, extended format, the bit width of L can be increased to $l = \lceil \log_2 (w+2) \rceil$ such that the value of q can be increased by one unit. In other words, by expanding the representation of q to take on values in the range $[0, w+1]$, the value of $q = w+1$ can be used to signal special conditions such as an error from division by zero. Some data can be associated with the special condition by storing the data within the value bits of A. This extension is not further discussed here.

14.6.3 Notation

To interpret the value $\mathcal{V}$ denoted by a bit representation V stored in a variable V, a base must be specified. Hence, to obtain the value stored in variable V, it is always qualified with a base v, as $\boxed{V}_v = Q \times v + R$, where Q and R are the quotient and remainder when $\mathcal{V}$ is divided by v, encoded in the bit representation V.

The notation $\mathcal{Q}(V)$ represents the value of the quotient part Q stored in the variable V, and $\mathcal{R}(V)$ denotes the remainder part R of V. They are independent of the base, so no base needs to be specified for V to obtain $\mathcal{Q}(V)$ and $\mathcal{R}(V)$.

Also, we will use the notation $\boxed{V}_{v:w}$ to denote the conversion of the value of V from a base v to a new base w. In other words, the bit representation of V is first viewed as representing the value $\mathcal{V} = \boxed{V}_v = Qv + R$. This value $\mathcal{V}$ is then converted to suit another base w such that $\mathcal{V} = \boxed{V}_w = Q'w + R'$ for some new values Q' and R'.

14.6.4 Signed Values and Modulo Adjustment

The representation format, the set of operations, and the result types are all defined only in terms of unsigned integers. Signed integers are handled by the application, independently from the reversible arithmetic here, by maintaining a sign bit to the representation and manipulating the sign bit separately as desired by the application.

Also, all arithmetic is with variable precision except in the case of the largest possible base value 2^w.

Modulo arithmetic is used if and only if the base S is equal to $W = 2^w$. The following modulo operation is applied in computing the addition operation $A \leftarrow A + B$ when $2^w \le A + B < 2^{w+1}$ (i.e., when the sum overflows w bits):

$$\boxed{W + x}_W \equiv \boxed{x}_W, \; 0 \le x < W.$$

Similarly, in computing the subtraction operation $A \leftarrow A - B$, the following modulo operation is applied when $-2^w < A - B < 0$ (i.e., when the difference is negative):

$$\boxed{0 - x}_W \equiv \boxed{W - x}_W, \; 0 < x \le W.$$

14.6.5 Backward Compatibility

The new representation can be viewed simply as a relaxation of the conventional representation, and hence is fully backward compatible with variables that have no specific qualification of a base. The relaxation is obtained as follows. Any integer $\mathcal{A}$ of the conventional representation can be viewed in the new representation as $\mathcal{A} = \boxed{A}_{2^w} = Q \times S + R$, where $Q = 0$, $R = \mathcal{A}$ and $S = 2^w$. In other words, the original value is equal to the remainder in the new representation, with quotient 0, and the divisor being sufficiently large to make even the largest possible value of $\mathcal{A}$ represented with w bits to be less than the divisor. Thus, by default, the new representation can be used with $q = 0$ and $S = 2^w$ for every conventional w-bit number.

The implementation of the new integer framework can be realized either natively in hardware or indirectly as an interface that sits over conventional arithmetic. The bit lengths required to support the alternative integer format are shown in Table 14.3 and Table 14.4 for different usable bit precisions. In the table, it is assumed that a hardware implementation supports conventional arithmetic for integers whose bit widths are powers of two (2^k for integer $k \geq 2$). With the native implementation approach, for example, a 32-bit integer interface may be implemented using 38-bit internal representations in hardware. Alternatively, the new framework can be realized as a software-level interface supporting 27-bit integers implemented over a conventional 32-bit hardware implementation.

TABLE 14.3: Internal Bit Widths for the Alternative Integer Representation Format Customized to Provide 2^k Bits of Usable Integer Precision

k	Usable Precision	Internal Bit Width
2	4	$4 + \lceil \log_2 (4+1) \rceil = 7$
3	8	$8 + \lceil \log_2 (8+1) \rceil = 12$
4	16	$16 + \lceil \log_2 (16+1) \rceil = 21$
5	32	$32 + \lceil \log_2 (32+1) \rceil = 38$
6	64	$64 + \lceil \log_2 (64+1) \rceil = 71$
7	128	$128 + \lceil \log_2 (128+1) \rceil = 136$

TABLE 14.4: Usable Integer Precision in the Implementation of the Alternative Integer Representation Format Using 2^k Bits

k	Internal Bit Width	Usable Precision
2	$4 = 2 + \lceil \log_2 (2+1) \rceil$	2
3	$8 = 5 + \lceil \log_2 (5+1) \rceil$	5
4	$16 = 12 + \lceil \log_2 (12+1) \rceil$	12
5	$32 = 27 + \lceil \log_2 (27+1) \rceil$	27
6	$64 = 58 + \lceil \log_2 (58+1) \rceil$	58
7	$128 = 121 + \lceil \log_2 (121+1) \rceil$	121

14.6.6 Computing $\hat{Q}$ and R for Base v

For any number $\mathcal{V}$ to be represented as $\boxed{V}_v$ and stored in variable V, first the values of Q and R are computed such that $\mathcal{V} = Qv + R$. Then, the value $\hat{Q}$ to be stored in the quotient of V is computed such that $Q = 2^q - 1 + \hat{Q}$ for the largest possible q, $0 \le q \le w$. The values R, $\hat{Q}$ and q thus computed are

stored in V as $\boxed{\begin{array}{c|c|c} R & \hat{Q} & q \end{array}}$.

14.6.7 Bit Representation Examples

- Suppose the 8-bit variable V holds the bit pattern $\boxed{01011\,|\,010}\,\boxed{0011}$. In this representation, $w = 8$, $W = 256$, $l = 4$, $q = 3$, $\hat{Q} = 2$, $Q = 9$, and $R = 11$. Hence, for a base value 15, $\boxed{V}_{15}$ gives $\mathcal{V} = 9 \times 15 + 11 = 146$. Similarly, $\boxed{V}_{12}$ gives $\mathcal{V} = 9 \times 12 + 11 = 119$, and $\boxed{V}_{20}$ gives $\mathcal{V} = 9 \times 20 + 11 = 191$.

- Suppose the value $\mathcal{V} = 102$ is stored in an 8-bit variable V. Because the default base for 8 bits is 256, the number is equal to $\boxed{V}_{256} = \boxed{102}_{256}$, and the bit representation is $\boxed{01100110}\,\boxed{0000}$ with $w = 8$, $W = 256$, $l = 4$, $q = 0$, $\hat{Q} = 0$, $Q = 0$, and $R = 102$. This value can be converted to another base, say, 10, as $\boxed{V}_{256:10}$ with bit representation $\boxed{00010\,|\,011}\,\boxed{0011}$ in which $q = 3$, $\hat{Q} = 3$, $Q = 10$, and $R = 2$. This in turn can be converted to another base, say, 2, as $\boxed{V}_{10:2}$ with bit representation $\boxed{000\,|\,10100}\,\boxed{0101}$ in which $q = 5$, $\hat{Q} = 20$, $Q = 51$, and $R = 0$.

14.6.8 Reversible Set of Arithmetic Operations

Table 14.5 shows the set of operations in the new integer framework, with the definitions of their forward and reverse computations. With these operations, every expression or sub-expression E appearing on the right-hand side (RHS) of an assignment $V \leftarrow RHS$ must be qualified with a base value e, as $\boxed{E}_e$. After the assignment, the variable on the left-hand side of the assignment will hold a value whose base is equal to the base of the final RHS value. Note also that for every expression E, the base e must always be non-zero. This ensures that multiplication by zero and division by zero are both disallowed in forward mode.

In Table 14.5, A' represents the post-operational value of A in the forward mode. In other words, both A and A' refer to the same variable A, where A refers to the value of A before a forward operation, and A' refers to the value of A after the forward operation on A has been applied. Similarly, C' represents the post-operational value of C in the forward mode.

TABLE 14.5: A New Set of Alternative Arithmetic Operations Reversible without Generating History

Typical	Alternative	
Forward	Forward	Reverse
$A' \leftarrow A + B$	$A' \leftarrow \left\lfloor \boxed{A}_{a:W} + \boxed{B}_{b:W} \right\rfloor_{W:a}$	$A \leftarrow \left\lfloor \boxed{A'}_{a:W} - \boxed{B}_{b:W} \right\rfloor_{W:a}$
$A' \leftarrow A - B$	$A' \leftarrow \left\lfloor \boxed{A}_{a:W} - \boxed{B}_{b:W} \right\rfloor_{W:a}$	$A \leftarrow \left\lfloor \boxed{A'}_{a:W} + \boxed{B}_{b:W} \right\rfloor_{W:a}$
$A' \leftarrow A \times B$	$A' \leftarrow \left\lfloor \boxed{A}_{a:W} \times \boxed{B}_{b:W} \right\rfloor_{W:\boxed{B}_b}$	$A \leftarrow \boxed{A'}_{1:a}$
$A' \leftarrow A/B$ $A' \leftarrow$ $(A \bmod B)$	$A' \leftarrow \boxed{A}_{a:\boxed{B}_b}$ $C' \leftarrow \left\lfloor \boxed{C}_{c:W} + \boxed{\mathcal{Q}(A)}_W \right\rfloor_{W:c}$ $C' \leftarrow \left\lfloor \boxed{C}_{c:W} + \boxed{\mathcal{R}(A)}_W \right\rfloor_{W:c}$	$A \leftarrow \boxed{A'}_{\boxed{B}_b:a}$ $C \leftarrow \left\lfloor \boxed{C'}_{c:W} - \boxed{\mathcal{Q}(A)}_W \right\rfloor_{W:c}$ $C \leftarrow \left\lfloor \boxed{C'}_{c:W} - \boxed{\mathcal{R}(A)}_W \right\rfloor_{W:c}$

■ **Accumulate** This is intended as a replacement for the conventional operation $A' \leftarrow A + B$ in which the value of a variable A is increased by the current amount in B.

In the forward addition operation

$$A' \leftarrow \left\lfloor \boxed{A}_{a:W} + \boxed{B}_{b:W} \right\rfloor_{W:a},$$

the variable A holding A is overwritten with $A + B$. The bit widths of A and B are assumed to be the same, equal to w, and hence $W = 2^w$. The bases a and b of A and B, respectively, are unchanged by the operation. This operation is reversed by recovering the pre-assignment value A from the post-assignment value A' via

$$A \leftarrow \left\lfloor \boxed{A'}_{a:W} - \boxed{B}_{b:W} \right\rfloor_{W:a}.$$

Example: Suppose the 8-bit variable A stored in base 25 has value $\mathcal{A} = 120 = 4 \times 25 + 20 = \boxed{A}_{25}$, and B stored in base 10 has value $\mathcal{B} = 95 = 9 \times 10 + 5 = \boxed{B}_{10}$. Then,

$$A \leftarrow A + B$$

is written as

$$A \leftarrow \left\lfloor \boxed{A}_{25:256} + \boxed{B}_{10:256} \right\rfloor_{256:25},$$

which overwrites A with

$$\boxed{\mathcal{A}+\mathcal{B}}_{256}\Big|_{25} = \boxed{0\times 256 + 215}_{256:25} = \boxed{8\times 25 + 15}_{25}$$

with bit representation $\boxed{01111\,|\,001\,|\,0011}$.

■ **Diminish** This is analogous to the **Accumulate** operation. In the forward addition operation

$$A' \leftarrow \boxed{\boxed{A}_{a:W} - \boxed{B}_{b:W}}_{W:a},$$

the variable A holding A is overwritten with $A - B$. The bases a and b of A and B, respectively, are unchanged by the operation. This operation is reversed by recovering the pre-assignment value A from the post-assignment value A' via

$$A \leftarrow \boxed{\boxed{A'}_{a:W} + \boxed{B}_{b:W}}_{W:a}.$$

■ **Scale** This is intended as a replacement for the conventional operation $A' \leftarrow A \times B$ in which the value of a variable A is scaled by B.

In the forward multiplication operation

$$A' \leftarrow \boxed{\boxed{A}_{a:W} \times \boxed{B}_{b:W}}_{W:\boxed{B}_b},$$

the variable A holding A is overwritten with the product $A \times B$. The bit width $2w$ of A must be twice the bit width w of B, thus making $W = 2^{2w}$. The pre-operational value A of A must not be greater than the maximum value $(2^w - 1)$ that can be assumed by B. Also, B must be non-zero. This operation is reversed by recovering the pre-assignment value A from the post-assignment value A' via

$$A \leftarrow \boxed{A'}_{1:a}.$$

The idea behind the reversal is that the product $A \times B$ is held in A as the quotient of the value when divided by the base $\boxed{B}_b$. This quotient is the same as the pre-assignment value of A. This value can be recovered by moving the quotient to the remainder, which can be accomplished by viewing the product as representing a value with base unity. Thus, $\boxed{A'}_1$ moves the value from the quotient place to the remainder modulo W.

Example: Suppose an 8-bit variable A stored in base 6 has value $\mathcal{A} = 15 = 2 \times 6 + 3 = \boxed{A}_6$, where $w_A = 2w = 8$, $W = 2^8 = 256$. Also, the

4-bit variable $\mathcal{B}$ stored in base 5 has value $\mathcal{B} = 13 = 2 \times 5 + 3 = \boxed{\mathcal{B}}_5$, where $w_B = w = 4$. Then,

$$A \leftarrow A \times B$$

is written as

$$A \leftarrow \boxed{\boxed{A}_{6:256} \times \boxed{B}_{5:256}}_{256:13},$$

which overwrites A with

$$\boxed{\mathcal{A} \times \mathcal{B}}_W \Big|_{\mathcal{B}} = \boxed{15 \times 13}_{256:13} = \boxed{15 \times 13 + 0}_{13}$$

with bit representation $\boxed{0000\ \vdots\ 0000\ \vdots\ 0100}$.

- **Shrink and Modulo** This is intended as a replacement for the set of conventional operations such as $A' \leftarrow A/B$ and $A' \leftarrow A \mod B$ that compute the quotient and remainders from the division of A by B. The new operation is a substitution for all those operations, but, unlike the old operations, remains reversible without the need for history.

In the forward division operation

$$A' \leftarrow \boxed{\boxed{A}_{a:W}}_{W:\boxed{B}_b},$$

which is simplified as

$$A' \leftarrow \boxed{A}_{a:\boxed{B}_b},$$

the internal representation of A is changed such that the current value A of A is divided by B, and the resulting quotient Q and remainder R are stored in an encoded form in A, where $A = Q \times B + R$. The bit widths of A and B are assumed to be the same, equal to w, making $W = 2^w$. This operation is reversed by recovering the pre-assignment value A from the post-assignment value A' via

$$A \leftarrow \boxed{\boxed{A'}_{\boxed{B}_b:W}}_{W:a},$$

which is simplified as

$$A \leftarrow \boxed{A'}_{\boxed{B}_b:a}.$$

After the division operation has been applied on A, it is possible to extract the quotient or the remainder values to store in another variable. The values are obtained using the operator $\mathcal{Q}(A)$ to access the quotient value and $\mathcal{R}(A)$ to access the remainder value. To reversibly record them in another variable C, the quotient or remainder is added to the value of C represented in base c.

Example: Suppose the 8-bit variable A stored in base 15 has value $\mathcal{A} = 100 = 6 \times 15 + 10 = \boxed{A}_{15}$, and B stored in base 3 has value $\mathcal{B} = 14 = 4 \times 3 + 2 = \boxed{B}_3$. Then,

$$A \leftarrow A/B \text{ or } A \leftarrow A\%B$$

is written as

$$A \leftarrow \boxed{A}_{15:\boxed{B}_3} \implies A \leftarrow \boxed{A}_{15:14},$$

which overwrites A with

$$\boxed{\mathcal{A}}_{15:14} = \boxed{6 \times 15 + 10}_{15:14} = \boxed{7 \times 14 + 2}_{14}$$

with bit representation $\boxed{00010\ 000\ 0011}$.

The operation is reversed via

$$A \leftarrow \boxed{A}_{\boxed{B}_3:15} \implies A \leftarrow \boxed{A}_{14:15},$$

which overwrites A with

$$\boxed{\mathcal{A}}_{14:15} = \boxed{7 \times 14 + 2}_{14:15} = \boxed{6 \times 15 + 10}_{15}$$

with bit representation $\boxed{001010\ 11\ 0010}$, which was the original format before the forward operation.

14.6.9 Combined Operation: A Simple Illustration

Consider a sequence of integer arithmetic operations in conventional programming that takes a temperature value C in Celsius as input and computes its equivalent value F in Fahrenheit as output. A forward-only computation pseudocode and its sample execution are shown in Algorithm 14.4. This code is not reversible because there is a loss of information (truncation of fraction) at line 4 due to integer division. While a Celsius value of 11 is converted to the Fahrenheit value 51, it fails to recover the Celsius value of 11 from the Fahrenheit value of 51 when executed in reverse without using a history log (from $F = 51$, it gives $C = 10$ instead of $C = 11$). Note that such a loss is unavoidable in traditional arithmetic even when floating point representation is used.

The same computation of the conversion can be performed in reversible fashion using the reversible integer operations previously described. The forward and reverse programs of the reversible version of the same code are shown in Algorithm 14.5.

Algorithm 14.4 Irreversible integer arithmetic for Celsius–Fahrenheit conversion

Forward-only Program	Execution Example
1: **read** C	1: $C \leftarrow 11$
2: $t \leftarrow C$	2: $t \leftarrow 11$
3: $t \leftarrow t \times 9$	3: $t \leftarrow 11 \times 9$
4: $t \leftarrow t/5$	4: $t \leftarrow 99/5$
5: $t' \leftarrow t + 32$	5: $t \leftarrow 19 + 32$
6: $F \leftarrow t$	6: $F \leftarrow 51$
7: **print** F	7: **print** 51

Algorithm 14.5 Reversible integer arithmetic for Celsius–Fahrenheit conversion

Forward Program	Execution Example
1: **read** $\boxed{C}_W$	1: $C \leftarrow \boxed{11}_W$
2: $t \leftarrow \boxed{C}_W$	2: $t \leftarrow \boxed{11}_W$
3: $t \leftarrow \boxed{\boxed{t}_W \times \boxed{9}_W}_{W:9}$	3: $t \leftarrow \boxed{11 \times 9 + 0}_9$
4: $t \leftarrow \boxed{\boxed{t}_9 / \boxed{5}_W}_5$	4: $t \leftarrow \boxed{19 \times 5 + 4}_5$
5: $t \leftarrow \boxed{\boxed{t}_5 + \boxed{32}_5}_5$	5: $t \leftarrow \boxed{51 \times 5 + 4}_5$
6: $F \leftarrow \boxed{t}_5$	6: $F \leftarrow \boxed{51 \times 5 + 4}_5$
7: **print** $\boxed{F}_5$	7: **print** $F = Q : 51, R : 4$

Reverse Program	Execution Example
1: **read** $\boxed{F}_5$	1: $F \leftarrow \boxed{51 \times 5 + 4}_5$
2: $t \leftarrow \boxed{F}_5$	2: $t \leftarrow \boxed{51 \times 5 + 4}_5$
3: $t \leftarrow \boxed{\boxed{t}_5 - \boxed{32}_5}_5$	3: $t \leftarrow \boxed{19 \times 5 + 4}_5$
4: $t \leftarrow \boxed{t}_{5:9}{}_9$	4: $t \leftarrow \boxed{11 \times 9 + 0}_9$
5: $t \leftarrow \boxed{t}_{1:W}$	5: $t \leftarrow \boxed{0 \times W + 11}_W$
6: $C \leftarrow \boxed{t}_W$	6: $C \leftarrow \boxed{11}_W$
7: **print** $\boxed{C}_W$	7: **print** $C = Q : 11, R : 0$

14.6.10   Reversal of Multiple Arithmetic Operations

Every individual operation in Table 14.5 is designed to be reversible. There is no information loss per se for any given operation as only one variable is modified in every operation, and the modification is constructive in nature (i.e., can be inverted). However, the overall reversibility of a set of operations must be carefully handled when the operations act on common variables as part of a complex control flow. An important source of irreversibility may arise in the case when two variables x and y are used in the program in two potentially conflicting ways: once as $\boxed{x}_y$ and later as $\boxed{y}_x$. This source of irreversibility can be avoided by using a programming convention that avoids reusing bases as normal numbers. Other similar conventions and guidelines will need to be evolved after additional experience is gained in programming with reversible arithmetic.

14.7   Reversal of Basic Arithmetic in Hardware

Reversible implementations of numerical computation in hardware is typically designed in a computing package called the Arithmetic Logic Unit (ALU), which is conceptually an addition to the central processing unit (CPU) of a computer. An ALU provides an interface to compute one or more functions on input bits. A sequence of *operand* bit vectors can be provided as input, and a corresponding sequence of *operator* bit vectors can be provided as control to choose the specific function to be applied on an input bit vector. The functions can be either logical operations (e.g., $A \wedge B$) or arithmetic operations (e.g., $A + B$).

The challenge in the design of a reversible ALU lies in the difficulty of optimizing the circuit to minimize the resources and computation time in supporting the set of all functions specified by the ALU control interface.

For example, a 2-input, n-bit, 4-function ALU takes two operands A and B, each n bits long, and implements four functions that can be selected using 2 control bits. In addition to the primary inputs, the circuit may, in general, require additional "clean" bits, which are essentially inputs with known values. The primary output consists of $2n$ bits comprising the result plus n bits holding a copy of one of the input operands. The output may additionally consist of "garbage" bits required by the circuitry. A "garbage-less" circuit may be defined as one that requires exactly the same number of bits $2n$ in the output as in the input, without any clean input bits and garbage output bits. The control bits that specify the choice of the function are passed through unchanged from the input to the output.

The metrics of interest in the design of reversible ALU are the logical width (which indicates whether any clean bits are required), the logical depth (which determines the latency of computation), and the number of gates (which determines the area, size, and resources needed to drive the circuit). The lower the value for each of the metrics, the better the design. The challenge, therefore, is to achieve reversibility while keeping the values for the metrics as low as possible.

There are essentially two different approaches to realizing reversibility of arithmetic operations in hardware: the compute-copy-uncompute approach and the reversible gates approach.

- The first approach uses the compute-copy-uncompute methodology in which results are first computed using a "normal" circuit dedicated to the forward computation of the arithmetic function. The results are copied to the output. Following that, a separate "inverse" circuit is used to "uncompute" the temporary bits created by the forward circuit and recover resources such as energy used in the forward execution. The circuits can be based on normal (irreversible) gates. An instance of this approach is the numerical and logical computing unit of the Pendulum computer [Vieri, 1995, Vieri et al., 1998, Vieri, 1999, Frank, 1999].

- The second approach more directly achieves reversibility by relying entirely on reversible gates. The desired function is realized as a circuit built from reversible gates, whose composition, by definition, ensures reversible computation between the circuit's aggregate input and output bits. An example of this approach is the ALU designed in [Thomsen et al., 2010].

A reversible ALU can be implemented in hardware using dedicated circuitry using classical gates, or using a Field Programmable Gate Array (FPGA), or using reversible gates including quantum gates. A few designs of the hardware implementation have appeared in the literature. In one of the designs of a reversible ALU using reversible logic gates [Guan et al., 2011], two n-bit input operands are accepted on which any one among 32 defined functions can be reversibly computed. The functions are split into two bins of 16 logical and 16 arithmetic functions on the input vectors. A total of $13n$ reversible gates are used in the circuit, and $12n + 1$ garbage output bits are produced. Also, sub-circuits such as ripple-carry adder designs are available that can be used as blocks within reversible ALUs [Desoete and Vos, 2002, Van Rentergem and De Vos, 2005, Thomsen and Axelsen, 2008]. In another ALU design, a garbage-less circuit is developed, albeit for fewer functions (five arithmetic-logical operations) using reversible gates [Thomsen et al., 2010].

14.8 Further Reading

Within the past century, humans have forged ahead quite far in the forward-only direction of computer-based arithmetic [Brent and Zimmermann, 2010]. For example, double-precision (64-bit) floating point operations are performed at unprecedented speeds of many billions of operations per processor per second. Yet, it is fair to say that there is a relatively limited understanding regarding reversibility of computer-based arithmetic.

Just as new electric vehicle technology requires a large amount of new research to reach the maturity of fast and economical gasoline vehicle technology, so too reversible numerical computation can be expected to take much additional research and development to become comparable or competitive with forward-only numerical computation that has already enjoyed many decades of concerted effort. It is clear that floating point computation in its current state of practice cannot be made reversible without significantly affecting the speed of computations. However, there is little by way of fundamental, theoretical impediments that prevents some new form of reversible arithmetic from reaching the same levels of performance as conventional floating point computation technologies. It is difficult to predict the amount of time it would take for reversible numerical computation to mature, but it is certain that it will supplement, if not supplant, forward-only numerical computation at some point in the future.

Reversible arithmetic and numerical algorithms have received attention, and some methods exist for realizing them at the hardware level using reversible circuits. Reversibility of integer multiplication has been addressed using reversible circuits [Kowada et al., 2006]. Reversible computation with space and time equal to irreversible computation has been proposed for modular exponentiation [Shor, 1997] and applied to a quantum factoring algorithm. The reversibility of modular integer arithmetic is realized in a modular arithmetic logic unit [Sakiyama et al., 2006].

The reversibility of numerical computation has received attention in the design of reversible languages:

- The **PSILISP** language supported a concatenation of two words of integers and the replacement of the two-word integer with integer quotient and integer remainder; this was part of a mechanism to provide a reversible division operation without history [Baker, 1992]. A multiplication is realized via the $\mathtt{mpy}(x,y,z)$ operation overwrites $x{:}y$ as $\boxed{x \cdot y \leftarrow y z + x}$, where $0 \leq x < z$. Here, $x{:}y$ represents that the value spans across the two variables. A division is performed using $\mathtt{div}(x,y,z)$, which views $x{:}y$ as a single value and splits that aggregate value into a quotient and remainder as $\boxed{x \leftarrow x : y \bmod z}$ and $\boxed{y \leftarrow (x : y - (x : y \bmod z))/z}$, where $0 \leq x < z$. These semantics ensure that the multiplication and division operators are in fact inverses of each other. Note that their counterparts in traditional forward-only languages are *not* invertible.

- In the **R** language, an interesting operator called the *fractional product* is defined in [Frank, 1999] with syntax of the form *integer* */ fraction* to help deal with numerical issues such as overflow conditions resulting from the product of two integers.

- Suppose that reversible programming languages prohibited multiplication by zero. This is the counterpart of prohibition of division by zero in conventional forward-only programming languages. Clearly, such prohibition, although helpful in making multiplication reversible, may seem draconian because the programming community has grown used to the validity of multiplication by zero. Nevertheless, it may be time to ponder why a luxury of multiplication by zero is so essential, and why (or at what cost to comfort and convenience) it can be avoided in common programming.

With regard to numerical error conditions, there is already a window of one-step backward execution possible now, due to the overflow and underflow information generated by the hardware. A generalization to $1 \leq n \leq n_{max}$ steps is needed for reversibility, so that the information about the most recent n error conditions is automatically maintained by the system, providing reversibility despite up to n abnormal conditions. Note that the moving window of overflow or underflow is needed to keep a success bit for every successful arithmetic operation and an error bit for every abnormal arithmetic operation encountered by the ALU at runtime.

There is clearly much more to be accomplished in the area of reversible numerical computation. In addition to the reversal of individual operations, it is necessary to investigate reversal properties of aggregates of operations, such as linear algebra operations, and understand the minimum history size requirements needed to reverse an aggregate operation such as matrix-matrix multiplication. International standards need to eventually evolve in order to make an impact on the software community. New programming language interfaces need to be explicitly defined that prohibit irreversible numerical operations from the outset. Algorithmic optimization of implementations is needed for specialized reversible frameworks, to eventually perform on par with the performance of conventional forward-only arithmetic that has been achieved from the immense amount of optimization work of the past decades. An important element of such optimization is the elimination of all irreversible non-determinism, such as the random noise introduced by intrinsic functions, numerical libraries, or the hardware. The notion of perfect reversibility may be relaxed with appropriate limits on errors, perhaps giving rise to new concepts such as approximate reversibility.

Chapter 15

Reversing a Sorting Procedure

An example of using a non-local view to achieve memory-efficient reversibility is the determination of a reversal method for sorting procedures. Consider any sorting procedure that takes N elements as input and sorts them *in place*. In any sorting procedure that moves the elements in the array to be sorted, the procedure uses assignment operations on the variables holding the elements. Each such assignment appears as an apparently destructive assignment when the assignment is viewed in isolation. In general, the number of destructive assignments can be as large as $O(N^2)$, depending on the sorting procedure. Such degeneration to destructive assignments results in a corresponding increase in the memory needed to store the history for reversal. Thus, a localized view of the individual instructions executed by a sorting procedure results in memory inefficiency of reversal. However, when the entire sorting procedure is examined with a non-local view, it is possible to reduce the amount of memory to $N \log_2 N$ bits. This minimal amount of memory needed to reverse a sorting procedure can be determined in three ways, each corresponding to a distinct reversal method.

1. **Saving the Input**: A copy of the input permutation P_I can be saved by keeping track of the identity of every element whenever the element is moved from one location to another in the array of numbers. This requires instrumenting the sorting procedure such that whenever an element is moved during the procedure, the identity $i \in [1 \ldots N]$ of the element is moved along with the element. The reversal is accomplished by simply rearranging the output elements with the inverse P_I^{-1} of the saved permutation P_I. The memory needed is equal to the memory for storing a permutation of N numbers, which is equal to the memory needed to store the numbers $1 \ldots N$. Because each number requires $\log_2 N$ bits, the total memory needed is $N \log_2 N$ bits.

2. **Permutation Ordinal Number**: The identity of the permutation P_I corresponding to I can be stored after sorting. The memory size needed to store P_I is equal to the number of bits needed to encode the integer that represents the ordinal number $D(P_I)$ of the input permutation in the list of all possible permutations $P_1, \ldots, P_{N!}$ generated by an enumeration procedure. This is equal to the number of bits needed to encode

any number $D \in [1 \ldots N!]$, which is equal to $\log_2 N!$. Using the approximation $N! \approx N^N$, we get the number of bits as $\log_2 N^N = N \log_2 N$.

3. **Saving the Control Flow Decisions**: If the procedure is a comparison-sort (i.e., sorting is based on comparing the values of pairs of numbers), then it is known that the lower bound on the number of comparisons C is equal to $N \log_2 N$. Any comparison sort will involve C conditional operations, one per comparison. To be able to reverse the sort procedure, it is sufficient to store the truth value of the condition as computed in the forward execution. Because there are at least C comparisons, the number of bits of information needed to remember the control flow for reversal is $C = N \log_2 N$ bits, which is the minimal number of bits needed.

Other sorting procedures that do not change the input but generate a separate copy of the input in sorted order, that is, those that do not sort in place, do not need any extra space other than the copy itself.

Chapter 16

Implementing Undo–Redo–Do

16.1   Application Model

Consider the application program, in general, as containing a set of objects $\{O_i\}$, potentially hierarchically organized as a directed-acyclic-graph. The program behavior contains a set of actions $\{A_j\}$ that operate on the objects. For example, in a word processing program, the objects could be graphical elements such as drawn shapes or text objects that together comprise the document, or the set of lines in a text editor [Leeman, 1986]. Modifications to the document are performed by action performed on an object (or a set of objects). On any object(s) **O** being acted upon, an action **A** could result in the modification of some of the objects' attributes **O.D[A]** that are specifically affected by the action **A**. For example, attributes such as color or size could be changed by a graphical or programmed menu action **A**, in which case the affected data **O.D[A]** would be the variables that store the color or size of the drawn object. In applications with no object delineation, a single global object can be trivially defined as one that encapsulates all the variables of the program.

This paradigm is typically instantiated at multiple levels within the same application. For example, in a Web browser, an undo list is maintained for the address entry field and another undo list is maintained for any text field displayed within an online form in a Web page being displayed within the browser. Moreover, the forward and backward movement across Web pages within the browser itself can be viewed as an undo-redo style of operation, and can even be implemented as such. The forward button maps to the Redo operation, and the backward button maps to the Undo operation, while the

Web address bar serves the Do operation. The object being modified by each is the current window's contents.

16.2 Data Structures

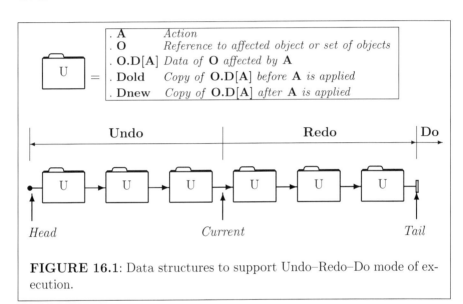

FIGURE 16.1: Data structures to support Undo–Redo–Do mode of execution.

The *Undo–Redo–Do* paradigm can be implemented using an undo data structure (Figure 16.1) with the Undo, Redo, and Do algorithms (Algorithm 16.1). The undo data structure is a doubly linked list of nodes that can be traversed in forward and backward directions. Each node in the list contains the information of an atomic action that can be done, undone or redone. The information in each list node includes (1) a reference to the set of objects, **O**, to which the action applies; (2) an action code, **A**, to indicate the type of action performed on **O**; and (3) the *before* and *after* copies of the specific portion of the objects' data that is modified by the action, named **Dold** and **Dnew**, respectively.

The undo list is logically divided into three segments, defined using three pointer variables: *head*, *current*, and *tail*. All nodes spanning from *head* to *current* represent actions that have already been previously done, and are ready to be undone if necessary. This segment of the list from *head* to *current* represents the *Undo* regime, which can be traversed backward to undo current state to a state in the past. Nodes from *current* to *tail* represent actions that have been done as well as undone. This segment of the list from *current* to *tail* represents the *Redo* regime, which can be traversed forward to reincorporate past actions that have been undone. New actions will be inserted starting at *current*. Before adding any new actions, all existing nodes in the *redo* segment are purged, because they have been marked as undone earlier and will be no longer relevant after a new path of actions is initiated starting from *current*. In other words, all existing actions that have been done and undone are deleted, thus making *current* become the new *tail*, and only then is a new action added. In a normal course of action, *current* would already be equal to *tail*, which results in all new actions become appended to the undo list, without any intermediate purging phase.

16.3 Algorithms

The Undo, Redo, and Do algorithms work in tandem as follows. For Undo, assuming that *current* has not reached *head*, the *current* node's action is undone by copying back the old data that was saved immediately prior to the action (described as part of the Do algorithm shortly). This copy operation restores the relevant portion of the object data, thus restoring the object to its overall state when the action is undone. For example, for a color change action, the graphical object's color would be restored to its prior color. Importantly, the undone node is not immediately deleted by the Undo algorithm, but it is retained in the undo list; only the *current* pointer is moved to the previous node in the list, denoting the new position for the next undo operation. This allows the possibility to redo the undone action at a later time without any additional external input.

The Redo algorithm works by "coasting forward" through the list from *current* to the next node. If *current* has not reached *tail*, it implies that the current node has been added by Do sometime earlier, but also undone sometime after that by an Undo operation. Such a node can be redone by simply copying back the post-action data into the object's relevant state. If *current* has reached *tail*, then a Redo at this point is interpreted as a request to apply the last node's action on a newly (currently) selected objects as targets. For example, if a color change action was applied to an object, and then a new object selected, a redo at this point would be treated as a request to apply the previous color change action to the newly selected object as well. Thus, when the Redo algorithm is invoked when *current* equals *tail*, it retrieves the most recent node's action and applies that action to the currently selected object(s). This, of course, is possible to be performed only if there is at least one node in the list (which is indicated by *current* being not equal to *head*).

The Do algorithm starts by purging all nodes, if any, from *current* to *tail*, because, once a new action is added at the current node, the previous nodes from *current* to *tail* can no longer be used for coast-forwarding in Redo.

After purge, *current* necessarily equals *tail*. A new node is then created and populated with the object and action information. The action is then applied, taking care to record the before and after snapshots of the relevant object data that is affected by the action. The new node is then added to the list.

Algorithm 16.1 Algorithms to support Undo–Redo–Do mode of execution

Undo	Redo

if *current* $\neq$ *head* **then**
    U $\leftarrow$ *current*
    U.**O**.D[U.**A**] $\Leftarrow$ U.**Dold**
    $--current$
end if
	if *current* $\neq$ *tail* **then**
	    U $\leftarrow$ *current*
	    U.**O**.D[U.**A**] $\Leftarrow$ U.**Dnew**
	    current++
	else if *current* $\neq$ *head* **then**
	    U $\leftarrow$ *current* -1
	    O $\leftarrow$ selected object(s)
	    Do(O,U.A)
	end if

Do(O,A)	

if *current* $\neq$ *tail* **then**
    Purge list: *current* to *tail*
end if
U $\leftarrow$ create a new undo node
U.**A** $\leftarrow$ **A**
U.**O** $\leftarrow$ **O**
U.**Dold** $\Leftarrow$ **O**.D[**A**]
Apply **A** on **O**
U.**Dnew** $\Leftarrow$ **O**.D[**A**]
Add U to list
current $\leftarrow$ *tail*
	<u>Notation</u>
	$x \leftarrow y$: Reference (shallow) copy
	$x \Leftarrow y$: Value (deep) copy
	$p - 1$: Previous node of p in list
	p++: Moves p to next in list
	$--p$: Moves p to previous in list

16.4 Deletions and Memory Reclamation

Because actions can be continually added to the list for all modifications to the objects, the list can grow indefinitely. The size of the undo list is only constrained by the amount of available memory. To reclaim memory from the undo list, nodes may be safely removed starting from *head* and up to *current*. Clearly, actions in the removed nodes will no longer be available for undoing later. Another option is to define a maximum list size and automatically purge the head node whenever a new action is added to the tail.

Deletion of objects is realized as an action on the container object of the objects being deleted, so that the container can retain the deleted object data in the **Dold** field of the corresponding node in the undo list.

16.5 Alternative Implementations

While the basic approach is the same in any implementation, different data structures can be used to store and retrieve the action list, and the method for recovering old values can be changed. For example, in place of the doubly linked list of actions, a pair of stacks can be used. Instead of saving object states to memory, state recreation may utilize reverse computation or re-computation.

16.5.1 Undo and Redo Stacks

Instead of a single undo list, the Undo–Redo–Do paradigm can be implemented using two stacks: (1) undo stack, and (2) redo stack. Essentially, when the undo list is split at the *current* pointer, the part from *head* to *current* can be implemented as a stack because operations on those nodes are performed in stack order (last in first out). Similarly, the part from *current+1* to *tail* becomes another stack, with the *tail* being at the base of the stack. Whenever a new action is performed, the Redo stack is flushed.

16.5.2 State Recreation via Reverse Computation

The reason that a copy of the old object data **O.D[A]** is saved before applying the action is that the data must be restored if the action is undone. However, memory and runtime overheads from the copying operation can be avoided if it is possible to recreate the overwritten state when the action is undone. This would require a procedure R to be defined that would be invoked to restore state at runtime whenever the action is done. If F is the original action procedure whose inverse is R, then, the signatures of the procedures are to be defined as $F(O, A_i, A_{i-1})$ and $R(O, A_i, A_{i-1})$, where A_i is the action being undone and A_{i-1} is its preceding action. The reverse procedure performs the inverse operation, A_i^{-1}, of A_i to restore the object. For example, in a graphical drawing application, if A_i is a "move" operation that translates the position of a graphical object by one grid point to the left, the inverse recreates the old state trivially by moving the object by one grid point to the right. This method of computing the inverse is, however, not useful if the action takes a long computation time (such as intensive image transformations, whose inverses also would incur inordinate computation expense).

Part IV

Hardware

Chapter 17

Reversible Logic Gates

17.1   Basic Concepts

Historically, logic gates have been used as building blocks for theoretical development of deterministic computation, and the assembled logic is realized in practice over physical processes such as composition of variable voltage levels in electronic circuit elements. However, much of the logic gate theory and design has focused on forward computation only, without explicit consideration of reversible execution. Here, we consider the reversibility aspect of traditional 2-bit gates and describe the concepts behind the development of reversible gates (which happen to require gates with at least 3 bit-wide inputs and outputs), and describe popular 3-bit reversible logic gates.

To be reversible, logic gates must have equal number of inputs and outputs—otherwise, some information in the input can be lost in the output and vice versa. In other words, for *reversibility*, the gates should possess bijection properties such that every input bit vector is uniquely mapped to an output bit vector and vice versa. This implies that reversible logic gates are confined to $w \times w$ permutation gates, where $w \geq 1$ is the number of input or output bits. For a given value of w, there are $2^w!$ (2^w factorial) distinct permutation gates that can be defined.

In addition to reversibility, reversible logic gates must provide *universality* properties, so that they can be used to synthesize any desired logic. Hence, only a subset of the permutation gates are useful as reversible logic gates in practice because some of them do not possess universality properties. For universality (being able to compute any given logic function), the triad of *and*, *or*, and *not* bit operations on 2-bit inputs forms the well-known necessary and sufficient operations to realize arbitrary computation. Equivalently, either a *nand* or *nor* operation also suffices, as any one of *nand* or *nor* can be used to realize the *and*, *or*, and *not* operations.

17.1.1 Inadequacy of 2-Bit Gates

Consider gates with $w = 2$, that is, the classical 2-bit gates. There are $2^2! = 4! = 24$ distinct gates that can be defined for 2-input, 2-output operations. Among these 24 gates, none of them qualify as reversible gates to implement *and* and *or* operations because all the gates that satisfy the functionality of *and* and *or* violate the bijection requirement between input and output bit vectors. This is easy to see by observing the requirements of *and* and *or* operations. The *and* operation requires mapping of the form $\{(\{(0,0),(0,1),(1,0)\} \rightarrow (0,\cdot)),(\{(1,1)\} \rightarrow (1,\cdot))\}$, where $\cdot$ denotes any bit value. Because this necessarily maps three input states to at most two output states, no permutation of a 2-input, 2-output gate can be reversible while computing the *and* operation. Similarly, the *or* operation requires mapping of the form $\{(\{(0,0)\} \rightarrow (0,\cdot)),(\{(0,1),(1,0),(1,1)\} \rightarrow (1,\cdot))\}$. Because this too necessarily maps three input states to at most two output states, no permutation of a 2-input, 2-output gate can be reversible while computing the *or* operation.

Thus, it is not possible to create reversible circuitry based solely on 2×2 gates. Hence, gates with $w \geq 3$ are needed to employ reversible logic gates. For $w = 3$, there are $2^3! = 8! = 40,320$ different permutation gates that satisfy the reversibility requirement.

17.1.2 w-Bit Gate Candidates, $w \geq 3$

Let us denote a permutation gate of w-wide input/output bit vectors by $_wG_P$, where, P is the permutation of input vectors. Let $_wG_{P^{-1}}^{-1}$ be its inverse permutation gate where $P^{-1}(P(x)) = x$. A subset of all possible $_wG_P$ that are their own self-inverses can be used as self-sufficient gates. In other words, if $P(P(x)) = x$ for all input bit vectors x, then $_wG_P = _wG_{P^{-1}}^{-1}$. Thus, two types of gates are possible:

Type 1 Gate pair (G, G^{-1}) such that $_wG_P = _wG_{P^{-1}}^{-1}$, where $P^{-1}(P(x)) = x$. P can be any permutation of the w-bit vector $[0, \ldots, w-1]$.

Type 2 Gate G such that $_wG_P = _wG_P^{-1}$, where $P(P(x)) = x$, and, therefore, $G = G^{-1}$. Clearly, this is a subset of Type 1.

TABLE 17.1: Number of Candidate Permutations for w-Bit Reversible Gates

Type	Number of Candidates
Type 1 (arbitrary permutations)	$2^w!$
Type 2 (self-inverse permutations)	R_{2^w}, where $R_N = R_{N-1} + (N-1)R_{N-2}$

In Type 1, there are $2^w!$ possible permutations of the 2^w possible bit vectors, and hence that many distinct gates that can be considered as candidates for reversible logic. However, only the gates that satisfy the universality requirement are actually eligible.

In Type 2, P is any permutation that is decomposable into 1-cycles and 2-cycles only, which makes the permutation a self-inverse. In other words, if $x[i]$ denotes the ith element of a bit vector x, then P only contains mappings in which $x[i] = P(x)[i]$ (which is a 1-cycle), or $x[i] \neq P(x)[i]$ and $x[i] = P(P(x))[i]$ (which represents a 2-cycle). Thus, Type 2 gates are those permutations in which either an element of the input is passed through unchanged or pairs of elements are swapped, or any combination of such pass-through and swap operations.

The number of possible permutations, R_N with $N = 2^w$, in Type 2 can be obtained from the recursion $R_N = R_{N-1} + (N-1)R_{N-2}$. This recursive equation is obtained using induction as follows: Given R_{N-1} for a Type 2 permutation of $N-1$ elements, a new element is now added. In permutations of N elements, the new element could either belong to a 1-cycle (i.e., map to itself) or belong to a 2-cycle (i.e., be mapped mutually to another element). In the former (1-cycle) case, the number of possible candidates for N elements remains the same as for $N-1$, as the new element does not interact with any of the existing $N-1$ elements. In the latter (2-cycle) case, the new element can be paired with any of the previous $N-1$ elements in $N-1$ ways, and leave $N-2$ elements to be permuted independently of the pair. Thus, the 2-cycle case gives $(N-1) \times R_{N-2}$ possibilities. The two cases (1-cycle and 2-cycle choices) are disjoint, and hence the total number of candidates is the sum of candidates in each case, which is $R_{N-1} + (N-1) \times R_{N-2}$. The recursion ends with $R_0 = R_1 = 1$.

17.2 3-Bit Reversible Gates

For any given w, we can obtain the number of distinct gates of Type 2 as R_N where $N = 2^w$. For $w = 3$, we get $N = 8$, and, $R_8 = N^4 - 6N^3 - 19N^2 + 136N - 132 = 764$. Thus, there are only 764 candidates for 3-bit reversible logic gates that are self-inverses. Among these gates, two of the most well-known Type 2 reversible gates that satisfy universality are the 3-bit versions of the Fredkin Gate and the Toffoli Gate.

17.3   Fredkin Gate

The Fredkin gate is a universal, reversible 3-bit gate that operates as a conditional router [Fredkin and Toffoli, 1982]. One of its inputs, viewed as a "control bit," determines whether the other two bits are passed through unchanged or swapped with each other. As shown in the input-output relation in Table 17.2, the input control bit always equals its corresponding output bit. The other two input bits are swapped at output if the control bit equals zero. The Fredkin gate is also known as a "controlled swap" (CSWAP) gate. The truth table of this gate is shown in Table 17.3.

TABLE 17.2: Input–Output Relations in a 3-Bit Fredkin Gate

Input	Output	Description
x_0	$y_0 = x_2 x_0 + \overline{x_2} x_1$	If x_2 is set, then $y_0 = x_0$ else $y_0 = x_1$
x_1	$y_1 = \overline{x_2} x_0 + x_2 x_1$	If x_2 is set, then $y_1 = x_1$ else $y_1 = x_0$
x_2	$y_2 = x_2$	Pass through unconditionally

TABLE 17.3: Truth Table for a 3-Bit Fredkin Gate

Input Bits			Output Bits			Permutation
x_0	x_1	x_2	y_0	y_1	y_2	
0	0	1	0	0	1	1-cycle
0	1	1	0	1	1	1-cycle
1	0	1	1	0	1	1-cycle
1	1	1	1	1	1	1-cycle
0	0	0	0	0	0	1-cycle
0	**1**	**0**	**1**	**0**	**0**	2-cycle ↕
1	**0**	**0**	**0**	**1**	**0**	
1	1	0	1	1	0	1-cycle

17.3.1   Reversibility

Reversibility of the Fredkin gate is easily verified. Visual inspection of the truth table verifies one-to-one and onto mapping between input and output bit vectors. When viewed as a permutation of input bit vectors, the permutation realized by the Fredkin gate contains six 1-cycles (elements unchanged in their position in the permutation) and a single 2-cycle (elements swapping places). This is indicated against each input vector in the truth table in Table 17.3.

17.3.2 Universality

Universality of the Fredkin gate is verified by construction to demonstrate how the gate can be used to realize the *and*, *or*, and *not* operations:

- The *and* operation can be realized by setting the x_1 input to 0, and using x_0 and x_2 as the input bits on which the *and* must be computed. This gives the desired *and* operation in the first output as $y_0 = x_0 \otimes x_2$, where $\otimes$ represents the *and* bit operator. This operation is shown in Figure 17.1.

- The *or* operation can be realized by setting the x_0 input to 1, and using x_1 and x_2 as the input bits on which the *or* must be computed. This gives the desired *or* operation in the first output as $y_0 = x_1 \oplus x_2$, where $\oplus$ represents the *or* bit operator. This operation is shown in Figure 17.2.

- The *not* operation can be realized by setting the x_0 input to 0 and the x_1 input to 1, and using x_2 as the input bit whose complement must be computed. This gives the desired *not* operation in the first output as $y_0 = \overline{x_2}$, where $\bar{x}$ represents the bit complement operator on x. This operation is shown in Figure 17.3.

For the Fredkin gate, the construction for the *not* operation can also be used to realize a fan-out operation: while the output bit y_0 gives the complement of x_2, the rest of the output bits y_1 and y_2 provide two copies of x_2, thus giving a fan-out of degree 2 for the input x_2.

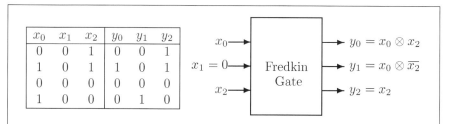

x_0	x_1	x_2	y_0	y_1	y_2
0	0	1	0	0	1
1	0	1	1	0	1
0	0	0	0	0	0
1	0	0	0	1	0

FIGURE 17.1: Realizing 2-bit *and* operation $\otimes$ with the Fredkin gate.

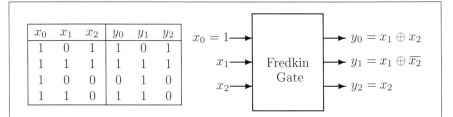

x_0	x_1	x_2	y_0	y_1	y_2
1	0	1	1	0	1
1	1	1	1	1	1
1	0	0	0	1	0
1	1	0	1	1	0

FIGURE 17.2: Realizing 2-bit *or* operation $\oplus$ with the Fredkin gate.

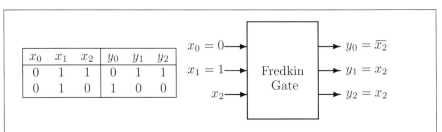

x_0	x_1	x_2	y_0	y_1	y_2
0	1	1	0	1	1
0	1	0	1	0	0

FIGURE 17.3: Realizing 1-bit *not* and fan-out with the Fredkin gate.

17.4 Toffoli Gate

Another well-known reversible logic gate is the Toffoli gate [Fredkin and Toffoli, 1982], also known as a "controlled controlled not" (CCNOT) gate. It can be expressed as a gate with w input bits, where $w \geq 2$. When $w = 2$, this gate is also sometimes referred to as the Feynman gate.

In a 3-bit Toffoli gate, all inputs are passed through unchanged except that, when the first two input bits are both 1, the third input bit is flipped on output.

17.4.1 Reversibility

Reversibility of the Toffoli gate is easily verified. Visual inspection of the truth table verifies the one-to-one and onto mapping between input and output bit vectors. When viewed as a permutation of input bit vectors, the permutation realized by the Toffoli gate contains six 1-cycles (elements unchanged in their position in the permutation), and a single 2-cycle (elements swapping places). This is indicated against each input vector in the truth table in Table 17.4.

TABLE 17.4: Truth Table for a 3-Bit Toffoli Gate

Input bits			Output bits			Permutation
x_0	x_1	x_2	y_0	y_1	y_2	
0	0	0	0	0	0	1-cycle
0	0	1	0	0	1	1-cycle
0	1	0	0	1	0	1-cycle
0	1	1	0	1	1	1-cycle
1	0	0	1	0	0	1-cycle
1	0	1	1	0	1	1-cycle
1	1	0	1	1	1	2-cycle ⇕
1	1	1	1	1	0	

17.4.2 Universality

Universality of the Toffoli gate is verified by construction of a 2-bit *nand* gate from a 3-bit Toffoli gate. Because *nand* gates (or *nor* gates) are universal gates, a Toffoli gate is also universal. Figure 17.4 shows the realization of a 2-bit *nand* operation from a 3-bit Toffoli gate. By fixing x_2 at 1, the output bit y_2 is observed to give the result of *nand* on inputs x_0 and x_1.

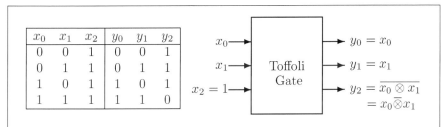

x_0	x_1	x_2	y_0	y_1	y_2
0	0	1	0	0	1
0	1	1	0	1	1
1	0	1	1	0	1
1	1	1	1	1	0

$x_0 \longrightarrow$ $y_0 = x_0$

$x_1 \longrightarrow$ Toffoli Gate $y_1 = x_1$

$x_2 = 1 \longrightarrow$ $y_2 = \overline{x_0 \otimes x_1}$
$= x_0 \overline{\otimes} x_1$

FIGURE 17.4: Realizing 2-bit *nand* operation $\otimes$ with the Toffoli gate.

17.4.3 Increasing the Width to w Bits

An expanded form of the Toffoli gate can be defined for any $w \geq 2$ bits, and it remains reversible for all w. However, it is not a universal gate when $w = 2$, but is universal for $w \geq 3$. The truth table for the w-bit form of the gate is shown in Table 17.5.

In the most general form, the w-bit Toffoli gate in fact accommodates an arbitrary Boolean function f of $w - 1$ bits for defining the remaining bit. The only requirement on f is that it is not always zero for all inputs. In the preceding introduction to the Toffoli gate, the bit-wise *and* function was assumed for f. A good overview and analysis is provided in [Rentergem et al.,

TABLE 17.5: Truth Table for a w-Bit Toffoli Gate

Input Bits					Output Bits					Permutation
x_0	$\ldots$	x_{w-2}	x_{w-1}	Property	y_0	$\ldots$	y_{w-2}	y_{w-1}	Property	
1	$\ldots$	1	**0**	$x_i = 1$ for all $0 \le i \le w-2$	1	$\ldots$	1	**1**	$y_i = 1$ for all $0 \le i \le w-2$	2-cycle $\Updownarrow$
1	$\ldots$	1	**1**		1	$\ldots$	1	**0**		
x_0	$\ldots$	x_{w-2}	x_{w-1}	$x_i \ne 1$ for some $0 \le i \le w-2$	x_0	$\ldots$	x_{w-2}	x_{w-1}	$y_i = x_i$ for all $0 \le i \le w-1$	1-cycle

2007]: with the generalized definitions, there are $w(2^{2^{w-1}} - 1)$ distinct Toffoli gates and $\frac{1}{2}w(w-1)(2^{2^{w-2}} - 1)$ distinct Fredkin gates.

17.5 Conservative Logic

In [Fredkin and Toffoli, 1982], the notion of *conservative logic* is presented as a way to ensure reversibility that can be implemented over reversible physical processes. In conservative logic, computation is viewed as simply *redirection* of signals rather than their *manipulation*. For a conservative gate, for every input bit vector, the number of 1's in the input must remain the same in the output vector. In that vein, no new bit values are created, and no bit value is destroyed. Any logic circuit whose truth table satisfies the bit-conservation can be seen as a router of bits from the input to the output.

The concept of conservative logic is separate from the concept of reversibility. There exist gates and their circuits that are reversible but not conservative. For example, circuits made from Toffoli gates are reversible but not necessarily bit conserving. In contrast, circuits made entirely of Fredkin gates are both bit conserving as well as reversible. A "billiard ball model" of computation [Fredkin and Toffoli, 1982] has been proposed as an example for physical realization of a reversible and conservative computation. The concept of conservative logic is motivated by the premise that it is easier to realize conservative logic over physical processes because the mapping from routing-based logic often maps naturally to the conservative processes prevalent in physical systems. Using an ideal system of billiard balls, a mechanical realization of a logical computation can be constructed. If the presence of a billiard ball represents the bit value 1, and absence represents 0, specific collision sequences may be carefully constructed using reflectors and precise angles of motion such that the desired overall logical computation is obtained as a sequence of collisions, with the number of balls being conserved in the system. In fact, several additional models, such as cellular automata and molecular chains, are studied under the generalization called *collision-based computing*. Conservative logic is discussed extensively in a book exclusively dedicated to collision-based computing [Adamatzky, 2002], with many interesting models, including one in which the evolution of the Game of Life simulation can be used as a substrate to realize Turing Machine model execution.

17.6   Synthesis of Reversible Circuits

The synthesis operation can be stated as follows: Given a truth table $\mathcal{T}_{orig}(w)$ of w input and output bits (possibly irreversible), find a composite circuit of reversible gates, each gate with width $w_g \leq 2w$, such that the synthesized circuit's truth table $\mathcal{T}_{syn}(w')$, $w' \geq w$, contains $\mathcal{T}_{orig}$ for some subset of input and output bits. Clearly, multiple synthesized circuits can be obtained for any given truth table, and hence the solution is not unique. Hence, some design criterion must be introduced to select an appropriate circuit from all the possibilities. The synthesis problem thus becomes that of finding a method that optimizes some type of cost measure associated with the synthesis, such as the number of gates or switching elements introduced into the circuit by the synthesis procedure. For example, the number of transistors required by silicon technologies varies for different gates. A good overview of the synthesis procedure with Toffoli or Fredkin gates is given in [Rentergem et al., 2007], which defines three types of cost measures (gate cost, switch cost, and quantum cost) and presents six different synthesis procedures that result in different values for each cost measure. Additional resources for developments in synthesis of reversible logic include [Al-Rabadi, 2004, Wille et al., 2008, Stankovi and Stojkovi, 2009, Drechsler and Wille, 2011].

A software kit for the design of reversible logic circuits is presented in [Soeken et al., 2012], and the state-of-the-art is described in [Drechsler and Wille, 2012]. Additional coverage on a broader classification of the types of reversible gates and their counts are provided in [Vos, 2010].

Physical circuit technology to support reversible computation is undergoing rapid progress, providing evidence that very low power chips based on adiabatic computing are forthcoming [Ren and Semenov, 2011].

Chapter 18

Reversible Instruction Set Architectures

18.1   Instruction Set Issues

At the lowest level of execution, the set of instructions provided by the computer architecture plays an important role in reversibility. Most conventional instruction sets are designed for forward-only execution. As a result, program execution cannot be reversed at the hardware level even if the program itself happens to be reversible at the software level. To enable reversibility natively, the instruction set itself must provide reversible primitives by design, to which compilers or program writers can target their code generation to develop reversible executables.

The core of an instruction set into which reversibility semantics must be infused is comprised of those instructions that provide arithmetic operations and the collection of jump instructions. Because most of the constructs in the higher-level programming languages are implemented in terms of arithmetic and jump instructions of the instruction set, reversible versions of the instruction set enable one to map reversible higher-level control flow to the reversible lower-level instructions, thereby preserving reversibility.

18.1.1   Instruction Set for Memory Operations

In conventional instruction sets, any variable can be overwritten with the value of any other variable or by a constant. However, because such blind overwriting can destroy information, the instruction set must provide either fully reversible updates or a clear distinction between reversible and irreversible updates. Instructions to exchange the values of two memory or register variables are self-inverses. Instructions that assign (i.e., copy) a value from a memory or register to another memory or register variable need special semantics. For example, such assignment (or *copy*) can be allowed if and only if the destination location contains zeros. A reverse instruction can easily restore the destination to zero, as no information has been lost. Similarly, newer instructions such as a *move* instruction (which zeros a location after copying the value to another location) can also provide reversible semantics by requiring the destination to be zero. Irreversible memory operations can also be supported by the hardware but their separation from the reversible versions help distinguish and isolate their effects (such as energy costs of the circuitry).

18.1.2   Instruction Set for Simple Arithmetic

In addition to the arithmetic used within the program itself, reversibility of arithmetic is also needed within the computer's hardware implementation for memory address computations such as offsets into arrays and other pointer arithmetic. However, the arithmetic within the user program requires more elaborate infrastructure to support a wider range of mathematical operations, including floating point arithmetic that is implemented separately in special hardware units (such as floating point units). Memory address computations defined within the instruction set require a simpler set of reversible arithmetic instructions such as addition and subtraction. Due to these considerations, the initial versions of reversible instruction sets in the literature have provided basic operators such as reversible addition and subtraction. Addition, for example, is defined as an accumulation operation with two operands $(A \leftarrow A + B)$ or overwriting operation with three operands $(C \leftarrow A + B)$. The accumulation instruction is undone by the inverse instruction $(A \leftarrow A - B)$, while the overwriting operation is undone by lossless erasure $(C \leftarrow C \text{ XOR } (A + B)$, where XOR is an exclusive-OR bit operation). Analogous reversible set of instructions are provided for bit-level operations of AND, OR, NOT, and EXCLUSIVE OR, such that irreversible binary operations require three operands while reversible unary or binary operations are defined with two operands. Some instruction sets make a distinction between arithmetic and logic operations; however, for simplicity, we view them as belonging to the same class or operations. While this is acceptable for integer arithmetic, the distinction becomes more pronounced when floating point arithmetic is considered.

18.1.3 Instruction Set for Jumps

Jump instructions in the hardware instruction set are the backbone for implementing the majority of complex control flow variants at the higher levels of programming. A conventional jump instruction, however, is irreversible in general because there is insufficient information at the jump's destination point. In reverse mode, when execution reaches an instruction labeled with a jump destination label, the control would not know whether to continue linearly back or jump back to the source of the jump instruction.

To make jumps reversible, the jump instruction is relaxed to a *jump to* instruction coupled with a new *jump from* instruction, and the semantics of the jump instruction are revised so that the destination of *jump to* is always a *jump from* instruction.

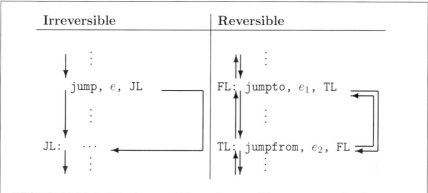

FIGURE 18.1: The irreversible and reversible constructs for jump instructions.

The conventional irreversible jump instruction and the modified reversible version are shown in Figure 18.1. The syntax jump, *e*, *label* indicates that the jump to the instruction labeled *label* is taken if and only if the condition *e* is satisfied. Typically, *e* refers to a single register or memory variable and checks if the variable holds a value that is greater than, equal to, or less than a constant or another register or memory location. The reversible version has similar syntax for the jump-to and jump-from instructions. Note that there are two, potentially different, expressions, e_1 and e_2, that operate in tandem for reversal. Just as the jump from FL to TL is taken in the forward mode if the condition e_1 is satisfied, the jump from TL to FL will be taken in the reverse mode if the condition e_2 is satisfied. In the forward mode, the jumpfrom instruction is a no-op; similarly, in the reverse mode, the jumpto instruction is a no-op. The programmer and/or the compiler are responsible to ensure that e_1 and e_2 are correctly paired so that this semantic of reversible jump instructions is maintained.

18.1.4   Implementation Considerations for Jumps

The hardware circuit implementation takes care of updating the program counter to reflect the new instruction addresses due to the jumps. These updates to the program counter are themselves made reversible using reversible arithmetic instructions. In the case of multiple jumps to the same destination label, disambiguation can be made by making use of a new variable with $\lceil \log J_{\mathrm{L}} \rceil$ bits, where J_{L} is the number of different `jumpto` instructions with the same destination label L. A dedicated variable for this purpose is assigned for each jump label.

Noting the symmetry of the `jumpto` and `jumpfrom` instructions, an implementation may merge both commands into one command, say, called `branch`, provided that there is a way to consult the current direction of control flow separately, say, via a special register bit that holds 0 for forward direction and 1 for reverse direction. Thus, a single branch instruction is sufficient to implement reversible branching if (1) the current direction of flow is available separately, and (2) the destination of every branch instruction is guaranteed to be another branch instruction.

18.2   Reversible PDP-10-Like Instruction Set Architecture

A popular instruction set known as the PDP-10 instruction set was originally designed for the conventional (forward-only) computer offered by the Digital Equipment Corporation (DEC), called the Programmed Data Processor (PDP) Model 10. In an early effort in the definition of a new Reversible Instruction Set Architecture (RISA) [Hall, 1994], an "isentropic instruction set" was modeled after the popular forward-only PDP-10 instruction set. The RISA included reversible memory operations, reversible arithmetic operations, and reversible control flow instructions, defined as follows.

18.2.1   Memory Operations

The instruction set includes memory manipulation primitives to operate on main memory locations and on registers. The operations allow swapping, copying, and movement of values between registers and/or memory locations. Operations for irreversible erasure of values are also provided. The memory operations are the EXCH, COPY, MOVE, and ERASE instructions, as listed in Table 18.1.

TABLE 18.1: Memory Operations in PDP-10-Like Reversible Instruction Set Architecture

Instruction	Description
EXCH *A*, *B*	Exchanges the values in *A* and *B*.
COPY *A*, *B*	Copies the value of *B* into *A*. The destination is required to be zero before the operation.
MOVE *A*, *B*	Moves the value of *B* into *A*. The destination is required to be zero before the operation, and the source is zeroed after the operation.
ERASE *A*	This is an irreversible operation that zeros the contents of the given location.
A and *B* are either registers or memory addresses potentially offset by a register	

18.2.2 Arithmetic

Reversible operations for simple arithmetic include the ADD, ADDM, and similar instructions, as listed in Table 18.2. Reversible operations such as subtraction SUB and bit-wise XOR can be used in two or three operand modes. Irreversible operations such as bitwise AND and OR can only be used in their three operand form (because the two operand form results in an irreversible update to one of the operands). The operations for ADD are illustrated in Table 18.2; other operations such as AND, OR, NOT, etc. are defined analogously.

TABLE 18.2: Simple Arithmetic Operations in PDP-10-Like Reversible Instruction Set Architecture

Instruction	Description
ADD *A*, *B*	*B* is added into *A*.
ADD *A*, *B*, *C*	*C*, assumed to be currently zero, is overwritten by *A+B*.
UNADD *A*, *B*	*B* is subtracted from *A*.
UNADD *A*, *B*, *C*	*C*, assumed to currently equal *A+B*, is zeroed by subtracting *A+B*.
A is a register, and *B* and *C* are registers or memory addresses offset by a register. *D* can also be a constant.	

18.2.3 Branches

Reversible jump operations include a family of jump instructions. The jump-to instruction has the following (simplified) syntax:

$$Jp \; reg, \; opnd, \; addr$$

where, p is one of GT, GE, EQ, LE, LT, or NE, and reg is a reference register, $opnd$ is a register or constant, and finally, $addr$ is the address label of the jump target. For example, the instruction

$$JLT \; I, \; 10, \; lbl$$

indicates that a jump to the statement labeled lbl must be made if the location I contains a value less than 10.

The jump-from peer of the jump-to instruction is called the *come from* instruction, obtained by replacing J with CF in the preceding syntax. Thus, the instruction

$$CFLT \; I, \; 0, \; lbl$$

indicates that, if the location I contains a value less than 10, a jump from the statement labeled lbl has been made to this instruction.

The jump instructions can be used to implement a reversible conditional statement as shown in Algorithm 18.1, and to implement a reversible loop statement as shown in Algorithm 18.2. In the reversible conditional statement, it is assumed that the variable var is not modified within the true or false branches. It is indeed possible to generate such a code if the source code originates from a reversible programming language. On the other hand, if the code originates from an irreversible language, a new (temporary) variable var will need to be introduced to hold the truth value beyond the conditional statement in the forward mode.

18.3   Pendulum Instruction Set Architecture

Another instruction set for reversible hardware is the Pendulum Instruction Set Architecture (PISA) [Vieri, 1995, Vieri et al., 1998, Vieri, 1999, Frank, 1999] originally designed for the Pendulum Chip aimed at very low-power computing. The basic approach for incorporating reversibility mechanisms is somewhat similar to that of the instruction set in the preceding section. However, it includes a more elaborate framework for additional classes of instructions such as reversible manipulations of bit vectors including bit rotations, expanded reversible arithmetic including multiplication and division defined for different bit precision widths, and reversible input and output semantics. Also included is a set of hardware-supported stacks to reverse otherwise irreversible instructions.

Algorithm 18.1 Realization of a reversible conditional statement using reversible jump instructions

Conditional Code	Instructions

if(*var*)
    true branch code
else
    false branch code
	`startlbl: JEQ` *var*`, 0, elselbl`
	⋮
	true branch instructions
	⋮
	`ifendlbl: J endlbl`
	`elselbl: CF startlbl`
	⋮
	false branch instructions
	⋮
	`endlbl: CFNE` *var*`, 0, ifendlbl`

Algorithm 18.2 Realization of a reversible looping statement using reversible jump instructions

Loop Code	Instructions

`for(`*var*`=0;` *var*`<`*num*`;` *var*`++)`
    loop body code
	`startlbl: CFGT` *var*`, 0, endlbl`
	⋮
	loop body instructions
	⋮
	`endlbl: JLT` *var*`,` *num*`, startlbl`

18.3.1 Memory Operations

An important design principle of this architecture is to provide all memory operations as exchange operations only. Also, common operations such as copying a value is achieved by ensuring that the destination holds a cleared state with zero and adding the copied value to the cleared state. Note that, for a memory location, a cleared state implies holding a zero value, but not vice versa.

18.3.2 Arithmetic

There is a large family of simple arithmetic and bit operations, such as AND (bitwise and), OR (bitwise or), XOR (bitwise exclusive or), NEG (two's complement negation), RL (rotate bits left), RR (rotate bits right), and so on. The reversible versions of unary and binary operators appear in one or two operand forms, with the one operand form specifying a single location to act as source and destination (e.g., NEG), while the two operand form uses both operands as sources and overwrites one of the operands as the destination (e.g., XOR). All the irreversible operations are offered in three operand form (e.g., AND).

18.3.3 Branches

A range of branching instruction codes is provided, including BEQ (branch if two specified registers hold equal values), BGEZ (branch if a specified register's value is greater than or equal to zero, BGTZ (branch if a specified register holds a value greater than zero), BLEZ (branch if a specified register's value is less than or equal to zero), BLTZ (branch if a specified register holds a value less than zero), BNE (branch if two specified registers' values are not equal), BRA (unconditional branch), and SWAPBR (swap the contents of a specified register with the contents of the branch register).

The reversibility of branching is supported differently from the PDP-10-like instructions of the preceding section. Here, a new computing model is assumed in which a special *direction bit* is supported by the architecture that controls the direction of control flow from any given state of the processor. The jump-from semantics are achieved by requiring that the destination of any branch is another branch instruction, coupled with the facility of reversing the flow of processor control via a special instruction called RBRA (toggle the processor direction bit).

18.3.4 Hardware Stacks

Two special stacks are defined in the original Pendulum architecture: a control stack and a data stack. The control stack, referred to as Program Counter Garbage Stack or PCGS, is intended to facilitate reversibility of jumps by pushing return addresses before making jumps, in a manner analogous to implementations of subroutine calls. This stack is especially useful in irreversible codes that cannot necessarily guarantee that the destination instruction is another branch instruction. The term "garbage" in the name of the stack relates to the view that the bits written to this stack are not necessarily reversibly erased, potentially contributing to higher energy use in their implementation due to blind bit erasure. Similarly, the data stack, referred to as `Datapath Garbage Stack` or `GS`, holds values of variables that are irreversibly overwritten.

18.3.5 Input/Output

Input to and output from the system are reversibly provided via a special *input* register with which other registers can interact via reversible bit operations. Specifically, reversible input is achieved via the `READ` instruction that specifies a register to be exclusively or'ed (`XOR`) with with the input register. The `SHOW` instruction copies a given register to the *output* register. The `EMIT` instruction moves a specified register to the output register.

Special `START` and `FINISH` instructions provide program initiation and termination facilities.

18.4 Hardware Interface to Reversible Memory

As an addition, or as an alternative, to processor-based reversible computing, reversibility may be supported by a suitably defined memory interface that can provide forward and backward execution semantics. The difference is that reversal is realized in terms of *previous values* of memory locations rather than *previous instructions* and their inverses operating on the memory locations. Hardware support for memory-based reversibility essentially comes in the form of a flexible versioning interface defined on memory units such as virtual memory pages.

One of the memory-based reversibility interfaces is the *Rollback Chip*, or RBC for short, that is envisioned as an auxiliary hardware unit that works in conjunction with traditional central processing unit and memory hardware [Fujimoto et al., 1992].

Imagine a versioning interface to a segment in main memory in which each page (or line) of memory may be individually and programmatically versioned at runtime. Memory is normally used by the program (without the concept of versioning, by default) for normal computation, but, to support reversibility in selected parts of the program, certain portions of memory may be programmatically declared to the RBC as requiring version tracking at runtime.

In addition to an initialization step, five types of operations are defined on

TABLE 18.3: Reversible Memory Interface of a Rollback Chip

Instruction	Description
READ A	Most recent (non-rolled-back) version of the memory location A is placed in A
WRITE A	The current value in the memory location A is recorded as the most recent version of A. If a value already exists for the current version in the history of A, that value is overwritten with this newly specified value
MARK	The version number of the reversible memory is incremented by one. For every location A in the memory, if one or more WRITE A operation(s) occurred between this MARK operation and the most recent MARK (or since initialization, if this is the first MARK), the latest of such written values persists as the value corresponding to this new version
ROLLBACK k	The value in every memory location A is rolled back k versions behind its current version
ADVANCE k	The k oldest versions of every memory location A are declared as discardable

the versioned memory, acting at the level of each memory location: (1) READ, (2) WRITE, (3) MARK, (4) ROLLBACK, and (5) ADVANCE. These operations are listed in Table 18.3.

The READ operation is used to get a read-only copy of the most recent, non-rolled-back version of a specified memory location. The WRITE operation is to push a new value to a specified memory location, which makes it the most recent version of that location. The MARK operation is used to specify the end of a version and start a new version. All writes from the previous mark (or initialization) to this mark are effectively coalesced, and the most recently written value becomes the lasting write between the two consecutive marks. Initialization of the versioned memory serves as a default, first mark on every versioned memory location. The ROLLBACK operation is used to restore any specified location to a version in its past, referred by its version number. The ADVANCE operation is a guarantee given to the versioned memory that no ROLLBACK operation will be invoked in the future to any version earlier than a specified version number. This helps the versioned memory hardware recover and reuse storage in which copies of earlier versions have been saved.

Note that the operations are provided as an *interface* only, making many optimizations possible in the implementation underneath the interface. For example, the ROLLBACK operation does not need to immediately perform any expensive memory copying, but instead perform just enough bookkeeping so as to retain the semantics for the next READ and WRITE operations. An implementation can also postpone internal updates and be able to merge the effects of multiple consecutive ROLLBACK operations. Effectively, the ROLLBACK instruction need only guarantee that the next READ operation on a memory location return the value corresponding to the correct version, which can be satisfied in a more lazy scheme that performs incremental updates to the internal directory system of the versioned memory.

18.5   Further Reading

This chapter focused on the fundamental ideas behind reversible instruction architectures. Additional detail can be found in the literature on the actual instruction sets with the precise bit layout of the machine opcodes and their operands, the circuits with clocking schemes and organization for energy reuse, and so on [Ressler, 1981, Hall, 1994, Younis and Knight, 1994]. For the PDP-10-like instruction set, see [Hall, 1994]; and for the Pendulum set, see [Vieri, 1995, Vieri et al., 1998, Frank, 1999, Vieri, 1999].

The issues in reversibly realizing the crucial "instruction fetch and decode" operation of a reversible processor, and specifically, how reversibility can be ensured by moving rather than copying opcodes from memory, are discussed in [Vieri et al., 1998].

Most authors bring attention to the distinction between a cleared state of a register (or memory location) and a zero state of that location. A cleared state is the state with a known value of zero. Although a value of zero may in fact be contained in a location at some point in time, it does not count as a cleared state because the circuit cannot be aware of its zeroed state and hence cannot exploit it to reuse energy for low-power operation [Vieri, 1995].

In the memory-based reversibility approach, additional hardware support for reversibility of execution traces is designed for debugging and other applications using an approach called the history cache [Sosic, 1994].

Overall, reversible computer architecture is in a relatively initial stage compared to the highly advanced state of conventional forward-only computer architectures. Many of modern techniques such as pipelining, etc., have not yet been incorporated in the context of reversible instruction sets, and much remains as open research.

Part V

Summary

Chapter 19

Future Directions

19.1 Phased Transition from Irreversible to Reversible

It is rather amazing that reversible computing naturally touches a rich variety of disparate areas, including thermodynamics, quantum physics, arrow of time, number theory, computability, randomness, abstract computing machines, nanoelectronics, languages, and compilers. Additional progress in reversible computing requires advances in both theory and practice, relying on many interdependent aspects that span all these areas.

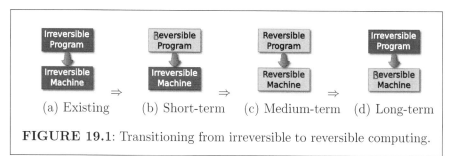

FIGURE 19.1: Transitioning from irreversible to reversible computing.

As reversible computing evolves in theory and software and hardware technologies, the transition from conventional computing may proceed in a series of steps, as illustrated in Figure 19.1. The existing state shown in Figure 19.1(a) corresponds to the most prevalent technologies of today, in which irreversible programs are executed on irreversible machines. A sudden transition from this current state to the ideal of a purely reversible computing world would require too gigantic a leap to be practically feasible. Instead, in the short term, a more effective intermediate step would be to achieve execution of new reversible programs over existing irreversible machines, shown in Figure 19.1(b). This includes the definition of novel and more powerful reversible languages for wider use, and an effective execution mechanism of the reversible programs

275

over traditional computing interfaces. Upon achieving such a capability, the focus may be shifted to replacing the irreversible machine with a reversible one, as shown in Figure 19.1(c). This includes possibilities such as the maturity of adiabatic circuit design and synthesis, and advanced CMOS technologies with circuits that are capable of transitions with predominantly recoverable energy. This stage of transition may also rely on quantum computing technologies that may provide machines of larger scale with many qubits, suitable for practical applications. The final achievement, shown in Figure 19.1(d), would be to be able to take any irreversible program and efficiently emulate it over a reversible machine. Clearly, this subsumes the capability of executing all the previous modes because reversible programs are a subset of irreversible programs. This is the idealistic final goal, in line with the vision of early works on adiabatic computing systems or reversible circuits.

19.2 Need for Additional Progress

Reversible computing represents, on many planes, a disruptive change. It brings changes in terms of theoretical ramifications of reversibility of computation in general. Computing characteristics such as performance per watt can become dramatically different due to the unique benefits offered by reversible hardware technologies. It is also disruptive due to the conflict it creates between the unique benefits that cannot be forsaken and the difficulty inherent in any paradigm shift. The difficulty arises due to the strikingly different, unfamiliar semantics imposed by reversible software systems. Consequently, just as it is true with any disruptive change, it faces severe hurdles for adoption.

Among the primary hurdles is the attitude of disbelief, perhaps due to its "spookiness." The research community has yet to overcome a disbelief that reversible computing is feasible, beneficial, and eventually inevitable as a natural evolution in computing. The second, and less subjective, hurdle is the comparison with the status quo, namely the immensely successful nature of irreversible computing versus the nascent state of reversible computing. Forward-only computing has advanced extremely far in terms of algorithms, data structures, computational complexity analyses, software infrastructures, software engineering methodologies, and so on, all built with irreversible computing as the underlying model. The forward-only notion has been ingrained in all programming minds. Also, the implementation technologies have advanced very far, making many applications easy, fast, and efficient. The barrier to reversible computing is the high bar set by the status quo with respect to ease of use, speed of computation, and intuition. Reversible computing must offer a clear and compelling case to warrant the large amount of initial pain in adoption. There are a few drivers at the moment that provide a compelling case to make the payoff work the effort.

- High Performance Computing (HPC) is an application area where reversible computing finds multiple uses:

  - Energy Efficiency: As parallel computers are populated with increasing numbers of processors, a critical limitation in scaling arises in the form of energy supply and dissipation levels. In some of the biggest supercomputing installations, input electric power has already reached nearly the highest levels (in the range of megawatts) that can be supplied to a single installation. Because the input power cannot be increased and dissipated heat cannot be decreased much more than current levels, future performance gains will ultimately rely on increasing the performance per watt using low-power computing technologies. In the direction of reduction of power usage, thermodynamic considerations eventually lead to the heat dissipation bounds imposed by logical reversibility. At this juncture, reductions in energy consumption can only be achieved by moving to reversible circuits.

  - Scalable Fault Tolerance: Due to a decrease of mean time between failure (MTBF) of system components with an increase in system size, hard and soft faults become much more frequent in large parallel computers. Reversible computing provides the least-cost solution for fault-tolerant execution realized via the ability to rollback processors as needed.

  - Scalable Debugging: In future HPC applications with very large concurrency levels (e.g., billions of threads of control flow or independent instruction streams), reversible computing is the least resource-hungry method of enabling debugging support for massively parallel runs.

- Quantum Computing is the impending computing paradigm of the future. Due to the fundamentally reversible nature of quantum computing operations, reversible computing must be accepted regardless of the challenges in programmability. In fact, research in reversible software for classical computing is being viewed by some as "practice" for impending adiabatic circuits and quantum computing hardware [Bennett, 2005].

In the near term, some challenges remain to be overcome. Robust, general-purpose reverse compilers need to be developed and refined to a state at which they are competitive with conventional compilers. All compilers, tools, and techniques need to be evaluated on real, large codes, and overall gains need to be evaluated. All new reversal schemes must be improved and refined so that they are in fact competitive with conventional (irreversible, sequential) execution.

In general, new standards need to be developed for reversible computer arithmetic, and new interfaces must be designed for reversible mathematical subroutines. Efficient algorithms are needed to achieve numerical reversibility, implementable at the hardware level for the fastest execution. In the process of enabling reversibility, a majority of conventional algorithms and data structures need to be revisited from their fundamentals. Their runtime complexity analyses must be redone, taking into account the high runtime cost of memory accesses and the ability to use reverse computation to recreate past state.

Additionally, some important, nontechnical challenges need to be addressed at some point before reversible computing can enter mainstream computing. Can we train a new breed of programmers who can become adept at writing programs using only reversible constructs? What would it take to teach computer science students to think in terms of reversible algorithms and reversible programming? How easy or difficult would it be to undertake software engineering on a new fundamental basis of reversible programming?

19.3   Outlook

Although there are several existing uses of reversible computing, it is possible that there are other, perhaps even more strongly tied, applications of reversible computing that will be discovered in the not-so-distant future. One observation that is certain is that backward execution is an undeniably natural peer to the conventional forward execution, and hence will eventually find its natural role in computing. How soon and at what pace it will unravel and fully blossom remains to be witnessed by this generation with intellectual curiosity, anticipation, hope, and excitement.

The strangeness of reversible computing is mainly due to our lack of experience with it.—*Henry Baker, 1992*

References

Andrew Adamatzky. *Collision-based Computing*. Springer, 2002. ISBN 978-1-85233-540-3.

Tankut Akgul and Vincent J. Mooney III. Assembly instruction level reverse execution for debugging. *ACM Trans. Softw. Eng. Methodol.*, 13(2):149–198, April 2004. ISSN 1049-331X.

Ian F. Akyildiz, Liang Chen, Samir Ranjan Das, Richard Fujimoto, and Richard F. Serfozo. Performance analysis of "time warp" with limited memory. In *SIGMETRICS*, pages 213–224, 1992.

A.N. Al-Rabadi. *Reversible Logic Synthesis: From Fundamentals to Quantum Computing*. Springer Series in Advanced Microelectronics. Prelim.Entry. 13, 13. Springer-Verlag, 2004. ISBN 9783540009351.

Eric Allen, David Chase, Christine Flood, Victor Luchangco, Jan-Willem Maessen, Sukyoung Ryu, and Guy L. Steele Jr. Project fortress: A multicore language for multicore processors. *Linux Magazine*, September 2007.

Eric Allen, David Chase, Hallett, Victor Luchangco, Jan-Willem Maessen, Sukyoung Ryu, Guy L. Steele, and Tobin-Hochstadt. The fortress language specification, version 1.0. Oracle, Inc., 2008, 2008.

James E. Archer, Jr., Richard Conway, and Fred B. Schneider. User recovery and reversal in interactive systems. *ACM Trans. Program. Lang. Syst.*, 6 (1):1–19, 1984. ISSN 0164-0925.

Holger Axelsen and Robert Glück. A simple and efficient universal reversible Turing machine. In Adrian-Horia Dediu, Shunsuke Inenaga, and Carlos Martn-Vide, editors, *Language and Automata Theory and Applications*, volume 6638 of *Lecture Notes in Computer Science*, pages 117–128. Springer Berlin/Heidelberg, 2011. ISBN 978-3-642-21253-6.

Johann Sebastian Bach. *The Art of the Fugue & A Musical Offering*. Dover Publications, 1747. ISBN 978-0486270067.

Mouad Bahi and Christine Eisenbeis. Register Reverse Rematerialization. Rapport de recherche (Research Report) inria-00607323, ALCHEMY - INRIA Saclay - Ile de France, July 2011.

Mouad Bahi and Christine Eisenbeis. Impact of reverse computing on information locality in register allocation for high performance computing. *International Journal of Parallel Programming*, pages 1–28, 2012. ISSN 0885-7458.

Henry G. Baker. Nreversal of fortune - the thermodynamics of garbage collection. In *Proceedings of the International Workshop on Memory Management*, IWMM '92, pages 507–524, London, UK, UK, 1992. Springer-Verlag. ISBN 3-540-55940-X.

Henry G. Baker. Thermodynamics and garbage collection. *SIGPLAN Not.*, 29(4):58–63, April 1994. ISSN 0362-1340.

Amir M. Ben-Amram. The Church-Turing thesis and its look-alikes. *SIGACT News*, 36(3):113–114, 2005.

C. H. Bennett. Logical reversibility of computation. *IBM J. Res. Dev.*, 17(6): 525–532, November 1973. ISSN 0018-8646.

Charles Bennett. Thermodynamics of computation. *International Journal of Physics*, 21:905–940, 1982.

Charles Bennett. Is information physical, or is physics informational? In *International Workshop on Reversible Computing (Special Session at ACM Computing Frontiers)*, 2005.

Charles H. Bennett. *On Random and Hard-to-describe Numbers*. IBM RC. IBM Thomas J.Watson Research Center, 1979.

Charles H. Bennett. Notes on the history of reversible computation. *IBM Journal of Research and Development*, 32(1):16–23, Jan. 1988. ISSN 0018-8646.

Charles H. Bennett. Notes on Landauer's principle, reversible computation, and Maxwell's demon. *Studies In History and Philosophy of Science Part B: Studies In History and Philosophy of Modern Physics*, 34(3):501–510, 2003. ISSN 1355-2198.

Charles H. Bennett, Ethan Bernstein, Gilles Brassard, and Umesh V. Vazirani. Strengths and Weaknesses of Quantum Computing. *SIAM Journal on Computing*, 26:1510–1523, 1997.

Phillip A. Berstein and Eric Newcomer. *Principles of Transaction Processing*. Morgan Kaufmann; 2 edition, 2009. ISBN 978-1558606234.

Antoine Bérut, Artak Arakelyan, Artyom Petrosyan, Sergio Ciliberto, Raoul Dillenschneider, and Eric Lutz. Experimental verification of Landauer's principle linking information and thermodynamics. *Nature*, 483:187–189, 2012.

P.G. Bishop. Using reversible computing to achieve fail-safety. In *Proceedings of the Eighth International Symposium On Software Reliability Engineering*, pages 182 –191, 2-5 1997.

Bitan Biswas and Rajiv Mall. Reverse execution of programs. *ACM SIGPLAN Notices*, 34(4):61–69, April 1999.

Ludwig Boltzmann. On the relation of a general mechanical theorem to the second law of thermodynamics. *Sitzungsberichte Akad. Wiss., Vienna*, 75: 67–73, 1877.

Bob Boothe. Efficient algorithms for bidirectional debugging. *SIGPLAN Not.*, 35(5):299–310, 2000.

Kevin J. Bowers, Edmond Chow, Huafeng Xu, Ron O. Dror, Michael P. Eastwood, Brent A. Gregersen, John L. Klepeis, Istvan Kolossvary, Mark A. Moraes, Federico D. Sacerdoti, John K. Salmon, Yibing Shan, and David E. Shaw. Scalable algorithms for molecular dynamics simulations on commodity clusters. In *Proceedings of the 2006 ACM/IEEE conference on Supercomputing*, SC '06, New York, NY, USA, 2006. ACM. ISBN 0-7695-2700-0.

Richard P. Brent and Paul Zimmermann. *Modern Computer Arithmetic*. Cambridge University Press, 2010. ISBN 978-0521194693.

J. S. Briggs. Generating reversible programs. *Software Practice Experience*, 17(7):439–454, July 1987. ISSN 0038-0644.

Leon Brillouin. *Science and Information Theory*. Academic Press, 2nd Edition, 1956. ISBN 978-0121349509.

Jeffrey Bub. Maxwell's demon and the thermodynamics of computation. *Studies In History and Philosophy of Science Part B: Studies In History and Philosophy of Modern Physics*, 32(4):569 – 579, 2001. ISSN 1355-2198.

Harry Buhrman, John Tromp, and Paul Vitanyi. Time and space bounds for reversible simulation. *Journal of Physics A: Mathematical and General*, 34: 6821–6830, 2001.

Christopher Carothers, Kalyan S. Perumalla, and Richard M. Fujimoto. Efficient optimistic parallel simulations using reverse computation. *ACM Transactions on Modeling and Computer Simulation*, 9(3):224–253, 1999.

A. K. Chandra. Efficient compilation of linear recursive programs. Technical Report Stanford Artificial Intelligence Project Memo AIM-167, STAN-CS-72-282, Computer Science Department, Stanford University, April 1972.

Ben Collins-Sussman, Brian W. Fitzpatrick, and C. Michael Pilato. *Version Control With Subversion - The Official Guide And Reference Manual*. CreateSpace Independent Publishing Platform, 2009. ISBN 978-1441437761.

282   *References*

C. J. Date. *An Introduction to Database Systems*. Addlson-Wesley, 8th edition, 2003. ISBN 978-0321197849.

Bart Desoete and Alexis De Vos. A reversible carry-look-ahead adder using control gates. *Integration*, 33(1-2):89–104, 2002.

Edsger W. Dijkstra. Program inversion. In *Program Construction, International Summer Schoo*, pages 54–57, London, UK, UK, 1979. Springer-Verlag. ISBN 3-540-09251-X.

R. Drechsler and R. Wille. From truth tables to programming languages: Progress in the design of reversible circuits. In *41st IEEE International Symposium on Multiple-Valued Logic (ISMVL)*, pages 78 –85, 2011.

Rolf Drechsler and Robert Wille. Reversible circuits: Recent accomplishments and future challenges for an emerging technology. In Hafizur Rahaman, Sanatan Chattopadhyay, and Santanu Chattopadhyay, editors, *Progress in VLSI Design and Test*, volume 7373 of *Lecture Notes in Computer Science*, pages 383–392. Springer Berlin Heidelberg, 2012. ISBN 978-3-642-31493-3.

Michel Dubois, Murali Annavaram, and Per Stenstrom. *Parallel Computer Organization and Design*. Cambridge University Press, 2012. ISBN 978-0521886758.

Margaret A. Ellis and Bjarne Stroustrup. *The annotated C++ reference manual*. Addison-Wesley Longman Publishing Co., Inc., Boston, MA, USA, 1990. ISBN 0-201-51459-1.

David Eppstein. A heuristic approach to program inversion. In *Proceedings of the 9th international joint conference on Artificial intelligence - Volume 1*, IJCAI'85, pages 219–221, San Francisco, CA, USA, 1985. Morgan Kaufmann Publishers Inc. ISBN 0-934613-02-8, 978-0-934-61302-6.

J. Fisher. Trace scheduling: A technique for global microcode compaction. *IEEE Transactions on Computers*, C-30(7):478–490, 1981.

Joseph A Fisher. *Very long instruction word architectures and the ELI-512*, volume 11. ACM, 1983.

Michael Frank. Introduction to reversible computing: Motivation, progress, and challenges. In *International Workshop on Reversible Computing (Special Session at ACM Computing Frontiers)*, 2005.

Michael P. Frank. *Reversibility for Efficient Computing*. PhD thesis, Massachusetts Institute of Technology, 1999.

Edward Fredkin and Tommaso Toffoli. Conservative logic. *International Journal of Theoretical Physics*, 21:219–253, 1982. ISSN 0020-7748.

James D. French. The false assumption underlying Berry's paradox. *The Journal of Symbolic Logic*, 53(4):1220–1223, 1988.

Richard M. Fujimoto. *Parallel and Distributed Simulation Systems*. Wiley-Interscience, 2000. ISBN 978-0471183839.

Richard M. Fujimoto, Jya-Jang Tsai, and Ganesh C. Gopalakrishnan. Design and evaluation of the rollback chip: Special purpose hardware for time warp. *IEEE Trans. Comput.*, 41(1):68–82, January 1992. ISSN 0018-9340.

James E. Gentle. *Random Number Generation and Monte Carlo Methods (Statistics and Computing)*. Springer, 2003. ISBN 978-0387001784.

Robert Glück and Masahiko Kawabe. A program inverter for a functional language with equality and constructors. In Atsushi Ohori, editor, *Programming Languages and Systems*, volume 2895 of *Lecture Notes in Computer Science*, pages 246–264. Springer Berlin / Heidelberg, 2003. ISBN 978-3-540-20536-4.

Robert Glück and Masahiko Kawabe. Derivation of deterministic inverse programs based on lr parsing. In Yukiyoshi Kameyama and Peter Stuckey, editors, *Functional and Logic Programming*, volume 2998 of *Lecture Notes in Computer Science*, pages 187–191. Springer Berlin/Heidelberg, 2004. ISBN 978-3-540-21402-1.

David Goldberg. What every computer scientist should know about floating-point arithmetic. *ACM Computing Surveys*, 23(1):5–48, 1991.

Herman H. Goldstine. *The Computer from Pascal to von Neumann*. Princeton University Press, 1980. ISBN 978-0691023670.

Georg A. Gottwald and Marcel Oliver. Boltzmann's dilemma: An introduction to statistical mechanics via the Kac ring. *SIAM Review*, 51(3):613–635, 2009.

Ananth Grama, George Karypis, Vipin Kumar, and Anshul Gupta. *Introduction to Parallel Computing*. Addison-Wesley, 2003. ISBN 978-0201648652.

Andreas Griewank and Andrea Walther. *Evaluating Derivatives : Principles and Techniques of Algorithmic Differentiation (Second Edition)*. Society for Industrial and Applied Mathematics, 2008. ISBN 0898716594.

Zhijin Guan, Wenjuan Li, Weiping Ding, Yueqin Hang, and Lihui Ni. An arithmetic logic unit design based on reversible logic gates. In *Communications, Computers and Signal Processing (PacRim), 2011 IEEE Pacific Rim Conference on*, pages 925 –931, 2011.

Storrs Hall. A reversible instruction set architecture and algorithms. In *Physics and Computation*, pages 128–134, 1994.

James Harland. Analysis of busy beaver machines via induction proofs. In *Proceedings of the Thirteenth Australasian Symposium on Theory of Computing - Volume 65*, CATS '07, pages 71–78. Australian Computer Society, Inc., 2007. ISBN 1-920-68246-5.

Berthold Hoffmann. Term rewriting with sharing and memoization. In *Proceedings of the Third International Conference on Algebraic and Logic Programming*, pages 128–142, London, UK, 1992. Springer-Verlag. ISBN 3-540-55873-X.

J.K. Hollingsworth. Critical path profiling of message passing and shared-memory programs. *IEEE Transactions on Parallel and Distributed Systems*, 9(10):1029–1040, 1998.

P. Hontalas, B. Beckman, M. DiLorento, L. Blume, P. Reiher, K. Sturdevant, L. V. Warren, J. Wedel, F. Wieland, and D. R. Jefferson. Performance of the colliding pucks simulation on the time warp operating system. In *Distributed Simulation*, 1989.

Reimann Hugo. *Text-book of simple and double counterpoint including imitation or canon*. Nabu Press, 1904). ISBN 978-1172452330.

Keith Hutchison. Is classical mechanics really time-reversible and deterministic? *The British Journal for the Philosophy of Science*, 44(2):307–323, 1993.

D. Jefferson and P. Reiher. Supercritical speedup [discrete event simulation]. In *Simulation Symposium, 1991., Proceedings of the 24th Annual*, pages 159–168, 1991.

David R. Jefferson. Virtual time. *ACM Transactions on Programming Languages and Systems*, 7(3):404–425, 1985.

David R. Jefferson, B. Beckman, F. Wieland, L. Blume, M. DiLorento, P. Hontalas, P. Reiher, K. Sturdevant, J. Tupman, J. Wedel, and H. Younger. The time warp operating systems. In *Symposium on Operating Systems Principles*, volume 21, pages 77–93, 1987.

Susmit Kumar Jha. *Towards Automated System Synthesis Using SCIDUCTION*. PhD thesis, EECS Department, University of California, Berkeley, Nov. 2011.

Mark Kac. Some remarks on the use of probability in classical statistical mechanics. *Académie Royale de Belgique, Bulletin de la Classe de Sciences*, 5(42):356–361, 1956.

Masahiko Kawabe and Robert Glück. The program inverter lrinv and its structure. In Manuel Hermenegildo and Daniel Cabeza, editors, *Practical*

Aspects of Declarative Languages, volume 3350 of *Lecture Notes in Computer Science*, pages 219–234. Springer Berlin/Heidelberg, 2005. ISBN 978-3-540-24362-5.

F. P. Kelly. *Reversibility and Stochastic Networks*. New: Cambridge University Press 2011; Original: Wiley, Chichester, 1979.

Youngbae Kim, J. S. Plank, and J. J. Dongarra. Fault tolerant matrix operations using checksum and reverse computation. In *Proceedings of the 6th Symposium on the Frontiers of Massively Parallel Computation*, FRONTIERS '96, pages 70–, Washington, DC, USA, 1996. IEEE Computer Society. ISBN 0-8186-7551-9.

Julian Knight, Steve Bull, and Gary Palmer. *Cricket For Dummies*. For Dummies, 2007. ISBN 978-0470034545.

Andrei N Kolmogorov. On tables of random numbers. *Sankhyā: The Indian Journal of Statistics, Series A*, 25(4):369–376, 1963.

Luis Antonio Brasil Kowada, Renato Portugal, and Celina Miraglia Herrera de Figueiredo. Reversible karatsuba's algorithm. 12(5):499–511, Jun 2006.

Rolf Landauer. Irreversibility and heat generation in the computing process. *IBM Journal of Research and Development*, 5(3):183–191, 1961.

J. L. Lebowitz. Time's arrow and Boltzmann's entropy. In *Physical Origins of Time Asymmetry*, pages 131–146. Cambridge University Press, 1994.

Y. Lecerf. Machines de Turing reversibles. insolubilite recursive en $n \in N$ de l'equation $u = \theta^n$, ou θ est un "isomorphisme de codes". In *Comptes Rendus Hebdomadaires des Séances de L'Académie des Sciences*, volume 257, pages 2597–2600. 1963.

Pierre L'Ecuyer and Terry H. Andres. A random number generator based on the combination of four lcgs. In *Mathematics and Computers in Simulation*, pages 99–107, 1997.

Pierre L'Ecuyer and Richard Simard. Testu01: A c library for empirical testing of random number generators. *ACM Trans. Math. Softw.*, 33(4), August 2007. ISSN 0098-3500.

Jooyong Lee. Dynamic reverse code generation for backward execution. *Electronic Notes Theoretical Computer Science*, 174(4):37–54, May 2007. ISSN 1571-0661.

George B. Leeman, Jr. A formal approach to undo operations in programming languages. *ACM Trans. Program. Lang. Syst.*, 8(1):50–87, January 1986. ISSN 0164-0925.

D. Levesque and L. Verlet. Molecular dynamics and time reversibility. *Journal of Statistical Physics*, 72:519–537, 1993. ISSN 0022-4715.

R. Levine and A. Sherman. A note on bennetts time-space tradeoff for reversible computation. *SIAM Journal on Computing*, 19(4):673–677, 1990.

Ming Li and Paul Vitanyi. Reversible simulation of irreversible computation. In *IEEE Conference on Computational Complexity (CCC)*, 1996.

Ming Li, John Tromp, and Paul Vitanyi. Reversible simulation of irreversible computation. *Physica D*, 120(1):168–176, 1998.

Michael Lienhardt, Ivan Lanese, Claudio Mezzina, and Jean-Bernard Stefani. A reversible abstract machine and its space overhead. In Holger Giese and Grigore Rosu, editors, *Formal Techniques for Distributed Systems*, volume 7273 of *Lecture Notes in Computer Science*, pages 1–17. Springer Berlin/Heidelberg, 2012. ISBN 978-3-642-30792-8.

Y.-B. Lin and B. R. Preiss. Optimal memory management for time warp parallel simulation. *ACM Transactions on Modeling and Computer Simulation*, 1(4), 1991.

Jon Loeliger and Matthew McCullough. *Version Control with Git: Powerful Tools and Techniques for Collaborative Software Development*. O'Reilly Media, 2012. ISBN 978-1449316389.

Chris Lutz and Howard Derby. Janus: A time-reversible language. Letter written in 1986 by authors, then at California Institute of Technology, to R. Landauer of IBM Inc.; work claimed to be dated in 1982, 1986.

D. Manivannan and Mukesh Singhal. A low-overhead recovery technique using quasi-synchronous checkpointing. In *Proc. IEEE Int. Conference on Distributed Computing Systems*, pages 100–107, 1996.

O. J. E. Maroney. The (absence of a) relationship between thermodynamic and logical reversibility. *ArXiv Physics e prints*, 2004.

O. J. E. Maroney. The (absence of a) relationship between thermodynamic and logical reversibility. *Studies in History and Philosophy of Science Part B: Studies in History and Philosophy of Modern Physics*, 36(2):355 – 374, 2005.

O. J. E. Maroney. Generalizing Landauer's Principle. *Phys. Rev. E*, 79(3): 031105, Mar 2009.

George Marsaglia. Diehard battery of tests of randomness, stat.fsu.edu/pub/diehard, 1995.

Armando B. Matos. Linear programs in a simple reversible language. *Theor. Comput. Sci.*, 290(3):2063–2074, January 2003. ISSN 0304-3975.

Makoto Matsumoto and Takuji Nishimura. Mersenne twister: a 623-dimensionally equidistributed uniform pseudo-random number generator. *ACM Trans. Model. Comput. Simul.*, 8(1):3–30, 1998. ISSN 1049-3301.

John McCarthy. The inversion of functions defined by Turing machines. *Automata Studies*, pages 177–181, 1956.

Pascal Michel. Small Turing machines and generalized busy beaver competition. *Theoretical Computer Science*, 326(1-3):45–56, 2004.

Shin-Cheng Mu, Zhenjiang Hu, and Masato Takeichi. An injective language for reversible computation. In Dexter Kozen, editor, *Mathematics of Program Construction*, volume 3125 of *Lecture Notes in Computer Science*, pages 289–313. Springer Berlin/Heidelberg, 2004. ISBN 978-3-540-22380-1.

John Von Neumann. *Theory of Self-Reproducing Automata*. University of Illinois Press, Champaign, IL, USA, 1966.

J. Orban and A. Bellemans. Velocity-inversion and irreversibility in a dilute gas of hard disks. *Physics Letters A*, 24(11):620 – 621, 1967. ISSN 0375-9601.

Bryan O'Sullivan. *Mercurial: The Definitive Guide*. O'Reilly Media, 2009. ISBN 978-0596800673.

Gheorghe Paun, Grzegorz Rozenberg, and Arto Salomaa. *DNA Computing: New Computing Paradigms*. Springer, 2006. ISBN 978-3540641964.

Kalyan Perumalla, Matthew Andrews, and Sandeep Bhatt. TED models for ATM internetworks. *SIGMETRICS Perform. Eval. Rev.*, 25(4):12–21, 1998.

Kalyan S. Perumalla. *Techniques for efficient parallel simulation and their application to large-scale telecommunication network models*. PhD thesis, Georgia Institute of Technology, 1999.

Kalyan S. Perumalla. Generating perfect reversals of simple linear codes. Technical Report GT-CC-03-04, College of Computing, Georgia Institute of Technology, 2003.

Kalyan S. Perumalla. Scaling time warp-based discrete event execution to 10^4 processors on the blue gene supercomputer. In *International Conference on Computing Frontiers*, pages 69–76, Ischia, Italy, 2007.

Kalyan S. Perumalla and Alfred J. Park. Reverse computation for rollback-based fault tolerance in large parallel systems. *Cluster Computing*, 16(2), 2013.

Kalyan S. Perumalla and Valdimir A. Protopopescu. Reversible simulations of elastic collisions. *ACM Transactions on Modeling and Computer Simulation*, 23(2), 2013.

Clifford A. Pickover. *The Mobius Strip: Dr. August Mobius's Marvelous Band in Mathematics, Games, Literature, Art, Technology, and Cosmology.* Basic Books, 2007. ISBN 978-1560259527.

Michel Raynal. *Concurrent Programming: Algorithms, Principles, and Foundations.* Springer, 2012. ISBN 978-3642320262.

Jie Ren and V.K. Semenov. Progress with physically and logically reversible superconducting digital circuits. *IEEE Transactions on Applied Superconductivity,* 21(3):780 –786, June 2011. ISSN 1051-8223.

Yvan Rentergem, Alexis Vos, and Koen Keyser. Six synthesis methods for reversible logic. *Open Systems & Information Dynamics,* 14:91–116, 2007. ISSN 1230-1612.

R.F. Resende and A. El Abbadi. On the serializability theorem for nested transactions. *Information Processing Letters,* 50(4):177 – 183, 1994. ISSN 0020-0190.

Andrew Lewis Ressler. The design of a conservative logic computer and a graphical editor simulator. Master's thesis, Massachusetts Institute of Technology, 1981.

Eleanor G. Rieffel and Wolfgang H. Polak. *Quantum Computing: A Gentle Introduction.* The MIT Press, 2011. ISBN 978-0262015066.

Brian W. Roberts. When we do (and do not) have a classical arrow of time. *Philosophy of Science,* 2012.

K. Sakiyama, B. Preneel, and I. Verbauwhede. A fast dual-field modular arithmetic logic unit and its hardware implementation. In *Circuits and Systems, 2006. ISCAS 2006. Proceedings. 2006 IEEE International Symposium,* pages 787–790, 2006.

John K. Salmon, Mark A. Moraes, Ron O. Dror, and David E. Shaw. Parallel random numbers: as easy as 1, 2, 3. In *Proceedings of 2011 International Conference for High Performance Computing, Networking, Storage and Analysis,* SC '11, pages 16:1–16:12, New York, NY, USA, 2011. ACM. ISBN 978-1-4503-0771-0.

A.N. Shiryayev. On tables of random numbers. In A.N. Shiryayev, editor, *Selected Works of A. N. Kolmogorov,* volume 27 of *Mathematics and Its Applications,* pages 176–183. Springer Netherlands, 1993. ISBN 978-90-481-8456-9.

Peter W. Shor. Polynomial-time algorithms for prime factorization and discrete logarithms on a quantum computer. *SIAM J. on Computing,* pages 1484–1509, 1997.

Mathias Soeken, Stefan Frehse, Robert Wille, and Rolf Drechsler. Revkit: An open source toolkit for the design of reversible circuits. In Alexis De Vos and Robert Wille, editors, *Reversible Computation*, volume 7165 of *Lecture Notes in Computer Science*, pages 64–76. Springer Berlin / Heidelberg, 2012. ISBN 978-3-642-29516-4.

Ray J. Solomonoff. A formal theory of inductive inference. Part i. *Information and Control*, 7(1):1–22, 1964a.

Ray J. Solomonoff. A formal theory of inductive inference. Part ii. *Information and Control*, 7(2):224–254, 1964b.

Ravishankar Somasundaram. *Git: Version control for everyone.* Packt Publishing, 2013. ISBN 978-1849517522.

Rok Sosic. History cache: Hardware support for reverse execution. *Computer Architecture News*, 22(5):11–18, December 1994.

Saurabh Srivastava, Sumit Gulwani, Swarat Chaudhuri, and Jeffrey S. Foster. Path-based inductive synthesis for program inversion. *SIGPLAN Notices*, 46(6):492–503, 2011a.

Saurabh Srivastava, Sumit Gulwani, Swarat Chaudhuri, and Jeffrey S. Foster. Path-based inductive synthesis for program inversion. In *Proceedings of the 32nd ACM SIGPLAN Conference on Programming Language Design and Implementation*, PLDI '11, pages 492–503, New York, NY, USA, 2011b. ACM. ISBN 978-1-4503-0663-8.

Milena Stankovi and Suzana Stojkovi. Reversible synthesis through shared functional decision diagrams. In Roberto Moreno-Daz, Franz Pichler, and Alexis Quesada-Arencibia, editors, *Computer Aided Systems Theory - EUROCAST 2009*, volume 5717 of *Lecture Notes in Computer Science*, pages 510–517. Springer Berlin/Heidelberg, 2009. ISBN 978-3-642-04771-8.

Vincent S. Steckline. Zermelo, Boltzmann, and the recurrence paradox. *American Journal of Physics*, 51(10):894–897, 1983.

Doron Swade. *The Difference Engine: Charles Babbage and the Quest to Build the First Computer.* Viking Adult, 2001. ISBN 978-0670910205.

Leó Szilárd. On entropy reduction in a thermodynamic system by inference by intelligent subjects. *Zeitschrift für Physik*, 53:840–856, 1929. (original title "Über die Entropieverminderung in einem thermodynamischen System bei Eingriffen intelligenter Wesen").

W. Teitelman. *INTERLISP Reference Manual.* Xerox PARC, Palo Alto, California, 1975.

W. Teitelman. Automated programming: The programmer's assistant. *Integrated Programming Environments*, 1984.

Dave Thomas and Andy Hunt. *Pragmatic Version Control Using CVS*. The Pragmatic Programmers, 2003. ISBN 978-0974514000.

Michael Thomsen and Holger Axelsen. Parallel optimization of a reversible (quantum) ripple-carry adder. In Cristian Calude, Jos Costa, Rudolf Freund, Marion Oswald, and Grzegorz Rozenberg, editors, *Unconventional Computing*, volume 5204 of *Lecture Notes in Computer Science*, pages 228–241. Springer Berlin/Heidelberg, 2008. ISBN 978-3-540-85193-6.

Michael Kirkedal Thomsen, Robert Glück, and Holger Bock Axelsen. Reversible arithmetic logic unit for quantum arithmetic. *Journal of Physics A: Mathematical and Theoretical*, 43(38):382002, 2010.

TOP500.org. Top500 supercomputer sites. http://www.top500.org, 2013. Accessed: 2013-05-01.

Yvan Van Rentergem and Alexis De Vos. Optimal design of a reversible full adder. *International Journal of Unconventional Computing*, 1:339–355, 2005.

Carlin Vieri, M. Josephine Ammer, Michael Frank, Norman Margolus, and Tom Knight. A fully reversible asymptotically zero energy microprocessor, 1998.

Carlin James Vieri. Pendulum: A reversible computer architecture. Master's thesis, Massachusetts Institute of Technology, 1995.

Carlin James Vieri. *Reversible computer engineering and architecture*. PhD thesis, Massachusetts Institute of Technology, 1999.

Alexis De Vos. *Reversible Computing*. Wiley-VCH, 2010. ISBN 978-3527409921.

G. Vulov, Cong Hou, R. Vuduc, R. Fujimoto, D. Quinlan, and D. Jefferson. The backstroke framework for source level reverse computation applied to parallel discrete event simulation. In *Simulation Conference (WSC), Proceedings of the 2011 Winter*, pages 2960–2974, 2011.

A. N. Whitehead and B. Russell. *Principia Mathematica, Vol. 1*. Cambridge University Press, London, 1925.

R. Wille, H.M. Le, G.W. Dueck, and D. Grosse. Quantified synthesis of reversible logic. In *Design, Automation and Test in Europe, 2008. DATE '08*, pages 1015–1020, 2008.

C.-Q. Yang and B.P. Miller. Critical path analysis for the execution of parallel and distributed programs. In *Distributed Computing Systems, 1988., 8th International Conference on*, pages 366–373, 1988.

Tetsuo Yokoyama. Reversible computation and reversible programming languages. *Electron. Notes Theor. Comput. Sci.*, 253(6):71–81, March 2010. ISSN 1571-0661.

Tetsuo Yokoyama and Robert Glück. A reversible programming language and its invertible self-interpreter. In *Proceedings of the 2007 ACM SIGPLAN symposium on Partial evaluation and semantics-based program manipulation*, PEPM '07, pages 144–153, New York, NY, USA, 2007. ACM. ISBN 978-1-59593-020-2.

Tetsuo Yokoyama, Holger Axelsen, and Robert Glück. Towards a reversible functional language. In Alexis De Vos and Robert Wille, editors, *Reversible Computation*, volume 7165 of *Lecture Notes in Computer Science*, pages 14–29. Springer Berlin/Heidelberg, 2012. ISBN 978-3-642-29516-4.

Saed G. Younis and Tom F. Knight. Asymptotically zero energy computing using split-level charge recovery logic. In *Proceedings of the International Workshop on Low Power Design*, pages 177–182, 1994.

Ernst Zermelo. On a theorem of dynamics and the mechanical theory of heat. *Annalen der Physik*, 57:485–494, 1896.

P. Zuliani. Logical reversibility. *IBM Journal of Research and Development*, 45(6):807 –818, nov. 2001. ISSN 0018-8646.

Index

Printed and bound by CPI Group (UK) Ltd, Croydon, CR0 4YY

23/10/2024

01777673-0011